Falling to Fly

Jason Pednault & Matheu DeSilva

Copyright © 2018 Matheu DeSilva & Jason Pednault
Cover Art by Mauro Reátegui Pérez
All rights reserved.
ISBN-10: 1722645768
ISBN-13: 978-1722645762
Published by Jungle Visions

In loving memory of Leslie Allison

CONTENTS

Prologue. Falling 1
I. *Wild Blood* 5
II. *Ruin* 13
III. *Seeking Refuge* 33
IV. *Presa River* 49
V. *Indigo Summers* 67
VI. *Gold Rushing* 89
VII. *King of the Worms* 107
VIII. *The Jaguar and the Monkey* 119
IX. *Fool's Gold* 141
X. *Jatanyacu* 159
XI. *Huaorani Prince* 171
XII. *Tequila* 199
XIII. *Peligro* 211
XIV. *Sisu* 223
XV. *Unlocking* 235
XVI. *Demons Meeting* 261
XVII. *Sanango* 291
XVIII. *Grolars* 307
Epilogue & Author's Notes 319

The following is a true story, based on actual events.

Some names of people and places have been changed or omitted to protect privacy.

Jason Pednault   Matheu DeSilva

Jason:
For Cliff and Jasmine – my past, my now, my future.

Matheu:
For my father, Ken French - the source of the wild in me.

Jason Pednault   Matheu DeSilva

Falling

Ever since I was a kid I always wanted to fly. Throughout my childhood, I would climb to the top of tall trees and look out, wishing I could take wing and soar, higher and higher, to travel great distances and see the world from afar; pristine, wild and beautiful.

As I grew, I was told repeatedly my dream wasn't possible. Humans, as a species, are not made to fly. Sure, I could catch a plane if I wanted, but under no circumstances would I be allowed the everyday exhilaration and freedom our feathered friends take for granted. Perhaps I should have accepted that and gone about my business, like a good little consumer. But I couldn't. I had to keep my dream alive.

Dreams are dangerous, however. They are both a gift and a curse. It's true our most magnificent accomplishments are born of dreams, but so are our most destructive - often without anybody realizing which is which until much later. Daydreams, night dreams, plans, plots and schemes become shelter, security, bridges, tunnels, weapons, wars, vehicles, smartphones, televisions - you get the picture. For good or ill, it is our ability to turn dreams into reality that sets us apart from the rest of the creatures who share our planet (or so I thought, but we'll get to that later).

The day my dream became real, I rose into a silvery sky, weightless and terrified. I was in the Ecuadorian Amazon, beaten and bloody, and flying over the edge of a cliff. Lush vegetation glistened and sparkled beneath me, and a dense canopy of rolling green stretched into the distance. Between and far below was what looked like a slash of a roiling cappuccino: the Jatanyacu river. Surging through the bottom of the precipice, it rumbled an insistent invitation to join the clatter of boulders being tossed downstream.

The two men responsible for my inaugural flight looked on, eyes impassive, their bare chests still heaving from the effort of murderous violence. I arced away from them, limbs flapping, clawing at thin air. The moment stretched into eternity, and I was almost able to convince myself I might stay aloft, held up by the whispery embrace of the breeze. Then gravity took over and I dropped like a stone.

Twenty feet down I smashed into a rocky outcrop and bounced off in a shower of mud and blood. The slight inclines and protrusions gouged into the cliff by erosion and landslides ensured every inch of my body took a hit as I spun, smacked, and skidded downward over rocks, shale, slick soil.

I fought to stay conscious and alert, twisting my head to avoid direct impacts to my fragile human skull. I was pretty sure the odds of surviving a hundred-and-fifty-foot fall into raging rapids were minimal but had no intention of reducing them further by closing my eyes and giving up. As I fell, an odd sense of calm washed over me. It was as if it wasn't me plummeting to my death at all, but somebody else, someone separate from me.

I grabbed at everything I could; stunted saplings, scratchy bushes, tufts of coarse grass, woody vines. Nothing held. I was falling too fast. The twigs and shrubs ripped away from the crumbling cliff-face like cobwebs, causing a constant avalanche of dirt and debris to rain down all around me.

Amid the confusion, I noticed a sharp movement below. I assumed it to be some kind of bird or cliff-dwelling animal. But it wasn't. It was the vines. I blinked, thinking it a trick of the eye, but still they kept on moving. Gathering into clumps, they snagged and buffeted me, first one way, then the other, slowing my fall and moving me further and further to the left.

Beyond the clusters of animated vines, I could see the near vertical cliff-face I was careening down was about to disappear. As much as the vines had helped, the drop-off still came up too fast and all too soon I was airborne again. At over sixty feet, this was the biggest drop so far. I knew the moment I hit solid earth again it would be game over.

Then something magical happened. Something I still can't explain. A thick curl of vine whipped out from the ledge above and wound itself around my left leg, yanking me back with an almighty crack. The pain in my hip joint was excruciating as I swung, like a pendulum, across and down. Ten feet above the bank, the vine unraveled and released me. The impact knocked the wind out of me, and set me tumbling and crashing down the final straight toward the river. My body should have been pulp and shattered bones by now. Instead, it just hurt like hell.

I stole a glance at the seething rapids and saw a lone tree in my path. Thick and sturdy, with dark, shaggy foliage, it jutted at an incongruous angle a few feet above river's edge.

I lunged at it and my forearms slapped against the smooth bark, while the rest of me kept going with the momentum of a freight train. Holding on to that tree took all the strength I had left. For a long and agonizing moment, I dangled, legs swinging above the churning water, hands slipping and scrabbling, until eventually I hauled myself up onto the trunk.

I was dazed, battered, bruised, but alive.

High on a combination of relief and adrenaline, I bent forward and kissed the tree. Turning, I reached out and squeezed the nearest coil of vine, wincing as a jolt of pain fizzed from my hip to my knee to my toes. Despite it all, I laughed. I had no idea what had just happened. Or how. It was so unreal. I stared at the sparse network of twisting vines snaking up the cliff-face and wondered if I could ever repay them for saving my life.

As if in response, all the leaves on the tree I was sitting on quivered and the trunk began to vibrate. There was a bone-shuddering smack somewhere beneath me and a boulder burst out through the branches in a spray of murky water. I watched it roll on downstream, my heart sinking further with every smash and splash. I felt like vomiting. I had survived the fall but was far from safe.

The river was rising, and monstrous rocks were pin-balling down it like pebbles. Above, the cliff-face loomed as sheer and insurmountable as a castle wall. As if my situation wasn't bad enough, I couldn't see out my right eye, my right knee wouldn't bend, my left femur felt like it was floating outside my hip socket, and my windpipe was so crushed and swollen I could barely breathe.

I was trapped and broken, and the only two people who knew where I was were the men who'd thrown me down here. I was finished. The bastards had killed me.

Or so I thought.

Jason Pednault   Matheu DeSilva

Wild Blood

"Hey Jason! Time to come down!"

Though still far off, the voice, and its accompanying whistle, were shrill and insistent. I didn't answer at first. There was no need. My sister always knew where to find me. I let the tiny wood beetle scurry across my palm and around the back of my hand, then returned it to the damp crevice where I'd discovered it, high up in the tree. Twigs cracked on the forest floor below. Kendra was getting closer.

"Coming," I hollered back.

There was a splintered stump off to the right overhead. I grabbed it and pulled myself away from my perch. Legs dangling, I dropped through the air, caught a lower branch, then swung forward and reached for the next. Hands and feet dancing back and forth, I continued my descent in a rustle of leaves.

Kendra was standing beside my bike when I seized the final branch and plunged out of the canopy. I hung there over the ground, my scraped-up knees in line with her eyes, and she gave my foot a gentle tug.

"Found you, monkey-boy."

I let go, landed, and smiled up at her. She grinned back and straightened my glasses.

"Come on," she said. "We don't want to miss our flight."

My bike was still wedged into the tree where I'd used it to reach the lower branches earlier. I pulled it free and wiped the muddy footprint from the saddle with my t-shirt, then pushed it up the trail behind my sister. Kendra's bike was stashed at the edge of the woods. We retrieved it and rode home together. She glided out in front - the spotless glittery purple paint on her bike sparkling in the sun - while I followed, peddling hard, mud spitting up from my tires, spattering my face and clothes.

"Look what the old mouser dragged in," said my grandma, the moment we came through our apartment door. "You need to wash, boy. And change your clothes. And what's your bike doing down there in the courtyard? I hope you'll be better behaved when you get to your father's house. We may not like the man, but we don't want him thinking your mom's not raising you right."

I glanced past Grandma into the TV room. Mom was watching

Jeopardy with Grandpa. She didn't see me. She was mouthing '*What is profit?*' to the man on the television. Grandpa noticed me though. He held my eye for a moment, then shook his head and looked away. He hadn't spoken to me in weeks. Not since I'd decided to watch Saturday morning cartoons instead of going fishing with him one wet and windy weekend. I was ten years old, and he would hold that grudge right up until he died, more than twenty years later.

Once I was scrubbed, fed and put in clean clothes, Mom piled me, Kendra and our bags into the car and took us to the airport. My dad had moved to Wyoming after my parents' divorce, and we were going to stay with him for the summer.

I loved my mom. My grandparents too, no matter how cold and distant they were. But I idolized my father. He was my hero, and I couldn't wait to see him.

When we reached the airport, I went over to the big window so I could watch the planes landing and taking off. Kendra sat nearby and read. Our mother went to speak to the lady standing at the counter. When the two of them had finished talking, the lady came over and beamed down at Kendra and me. There were gold wings on her tight blue jacket and gold piping along her sleeves and lapels.

"How y'all doing?" she asked. Her teeth were white and straight. Her lipstick bubble-gum pink and shiny. "Jason, Kendra, I hear y'all are going to be flying with me this afternoon?"

Kendra nodded. I just stared.

"Well, now don't you worry about a thing. I'm Lorraine. And I'm going to be taking good care of you. You need anything, you ask me, okay?"

Lorraine's voice went up at the end of almost every sentence, making it appear as though she was asking a series of nonsensical questions. But she was so pretty and kind that I didn't care.

Once we'd said goodbye to Mom, Lorraine took Kendra and me by the hand and led us down the tunnel toward the plane. Halfway down, I glanced back. Mom was crying and waving. I waved back. Lorraine gave my hand a squeeze.

"You're going to be just fine, Cutey-pie," she said, or asked. I nodded and let her lead me on.

Lorraine brought us crayons and some paper with cartoon characters printed on them. Then she went to help another lady lift an old man out of a wheelchair and into the seat in front of us. Once the old man was settled, the rest of the passengers boarded. I watched them pass while Kendra read. Every now and then Lorraine would catch my eye and wink.

When we landed in Wyoming, Lorraine took us up the tunnel and into the arrivals lounge. A crowd of people waited, held back by a thick navy-blue rope strung between waist-high chrome columns.

I saw him first. Standing between two tall men in shiny suits and cowboy hats. He was looking straight at me and grinning like a fiend. *Wild blood* - that's what my grandparents said about my dad. And me, whenever I did anything wrong. He looked like a bear or a wolf-man. Even his scuffed brown jacket had fur on the collar. I forgot all about Lorraine and ran to him.

He lifted me into his arms, and then hoisted Kendra up beside me so he could hug us both at the same time. We were laughing and talking all at once. Dad's eyes glistened as he looked back and forth between us, telling us how much we'd grown and how much he'd missed us. I felt like I was going to cry so I buried my face into his neck and breathed in his smell; a mix of pine resin, sweat and beer.

"You remember Mary, right?" he said when he lowered us to the ground.

"Yes, of course! Hello Mary," said Kendra, stepping forward to give my dad's girlfriend a hug.

Mary was a pretty lady with a mass of curly brown hair. We'd met her when my dad came to see us on his motorcycle earlier in the year. She'd clung to his back when he arrived, and to his arm for the rest of the visit. I gave her one of my grandfather's curt nods then turned back to my dad.

"Are we going camping and fishing, Dad?"

He chuckled. "I promised, didn't I? As a matter of fact, we're leaving right now. I brought your birthday present with me. I was going to give it to you when I last visited, but figured it would be better to wait until now. Not only that, I have a cool surprise for you both. Something you're going to love."

He took mine and Kendra's hands, and we headed for the parking lot. It was late afternoon but the sun was still high above the horizon. I scanned the cars until I spotted his blue van. There was something big and yellow tied to the roof.

"A boat!" I said, bouncing up and down with excitement.

"Yep, that's right. It's adventure time, kids!"

The boat's outboard engine was in the back of the van with all our camping gear and fishing rods. The stink of diesel and river-slime made me queasy, and I spent the next hour looking straight ahead at the road so I wouldn't be sick. When we pulled onto a patch of gravel at the edge of a small lake, my dad wound down his window and leaned back to yank open the sliding door next to me. Fresh air rushed in and I felt instantly better.

"Break time. Kendra, honey, can you get the sandwiches from the cooler behind the bench? Grab a couple of beers for Mary and me, too, and some sodas for you and your brother."

Kendra handed out the sandwiches, and we stayed in the van gazing at the lake while we ate. I didn't notice the mustard until my second bite.

"Urgh, Dad! There's mustard in my sandwich!" I said, spitting the half-chewed mouthful out the door.

"Mine too," he said cheerfully, and everyone laughed.

"I'm sorry," said Mary. "I put mustard in all of them. Cliff didn't tell me you don't like it."

"Come on Jason, just eat it," said Dad. "She didn't have to make you anything, you know."

"Everyone knows I hate mustard!"

My dad gave me a hard look and I shut up. He didn't look away until I took a small bite and chewed. All I could taste now was the hot, vinegary spice. It obliterated all other flavors and made my nostrils itch and my eyes water. By the time I finished my sandwich I had tears streaming down my cheeks. When we headed back out on the road, I scowled at the back of Mary's head until, eventually, I fell asleep.

Camping trips used to be just me and Dad and Kendra. Mom never wanted to go with us. Mary was different; she loved the outdoor life. Which meant she and my father were more well-suited to each other than my parents had ever been. The way I saw it, Mary's arrival in my dad's life had destroyed any chance my parents had of getting back together.

The van shuddered and shook as it trundled up the rutted track to our campsite. I opened my eyes and looked up through the window at the soft indigo sky. Sprays of green leaves swept by, sparse at first, then becoming a dense tunnel of dappled light and deep shadows. I took off my glasses and cleaned their smudged lenses before sitting up. It occurred to me I'd already gotten used to the smell of the boat engine.

It no longer made me nauseous at all. I glanced at Mary, wondering if it might ever be possible to get used to her too.

My dad caught my eye in the rearview mirror. He smiled. "Nearly there, son."

I smiled back. The vast lake glinted everywhere between the trees now, and above me the fishing poles rattled in their racks.

Our campsite was in a clearing on the lakeshore, with the skeleton of an old jetty stretching out across the water in front. There were tall trees all around. Before leaving, I intended to climb every one of them.

The next morning, we woke early and got the boat out on the lake. It took a while to load everything up and get the motor running, but as soon as it was all ready, we set off across the still water for a blissful day of fishing. I was particularly happy, because Mary had decided to stay behind. She told us she had a book she wanted to finish, and waved us off with it wedged under her arm.

Once we reached the middle of the lake, my dad cut the engine and let the boat drift a while. He asked us about school and how things were back in Arkansas, while unpacking our gear and handing out our fishing poles. The one he gave me was new, and a real beauty. Metallic blue and lightweight, with an open reel. Up until then I'd always used a closed-face reel like most kids. This was the first time I'd been entrusted with an actual grown-up rod, and I puffed up with happiness and pride.

"Happy Birthday, Jason," said Dad, nodding at the fishing rod. "Sorry I wasn't there on the day, but I hope you like your present."

I told him I did, although I'm sure the look on my face already said it all. We spent some time practicing casting and reeling, until I got used to the feel of my new rod. Once I was ready, we puttered across the lake, stopping again at a place Dad said felt lucky.

"Let's hope it's lucky for us and not the fish," said Kendra.

Dad smiled and nodded while deftly baiting our hooks. I was the first to get a bite. Dad helped me reel it in, as I wasn't quite used to my new rod yet. It turned out to be a tiny lake trout, no more than a few inches long. I was disappointed but still wanted to keep it. Dad insisted we throw it back.

"But it's a lucky fish," I protested. "The first one of the day. The first with my new rod."

"It *is* a lucky fish," my dad replied. "That's why we're going to set him free. Maybe you'll meet again in a few years once you've both had time to get a little bigger. At the moment, though, he's just a kid like you. How would you feel if you were seized upon and eaten up before you had a chance to grow?"

"Not too good," I admitted, trying not to look at the tiny trout writhing and gasping on my hook.

We returned the fish to the lake and I cast off again. The spot soon proved to be as lucky as Dad had predicted. By late morning, we had all caught a good number of decent-sized fish – mostly perch and trout - and were ready to head back to camp for lunch. Before we could, however, my line went taut again. I was getting a feel for the new rod and could tell I'd snagged a big one.

The line spun out in a whir, and I gripped the rod with both hands. I looked to my dad, eyes wide, and tried to pass it to him.

"It's okay Jason, you've got this. Keep it up and start reeling her in. Nice and easy."

I caught the reel just before the line ran out and began teasing it back in. Every now and then my dad would use the back of his hand to raise the tip of the rod if I let it drop, but otherwise it was just me against the fish. Almost in a trance, I began winding it in. Not too fast, not too slow. Drops of sweat stung my eyes. I blinked hard, but didn't lose concentration.

"You're doing great, son. He's yours!" My father's voice sounded a million miles away.

A splash and a flash of color; the fish surfaced twenty feet away. I heard Kendra whistle, low and long. It would be the biggest fish either of us had ever caught. If I could land it. The monstrous trout dived away and I let the line out a little. Feeling for when the fish slowed, I gently raised the rod and coaxed it in again. It broke the surface a second time. Leaping toward us before twisting and turning away. I thought I was going to lose him. But somehow I managed to keep him snagged. The sky darkened for a split-second, as if a lone cloud had raced out in front of the sun. The hairs on the back of my neck prickled. Something else was happening.

"Look," said Kendra, pointing up.

I ignored her and kept my attention on my line, the surface of the lake. Then I saw it. A rush of brown and white feathers, diving for the lake. At the last moment, the bald eagle, for that's what it was, leaned back. Enormous wings buffeting the air, while its feet swung forward, glancing off the surface of the water. In one fluid motion, it seized and shifted, thrusting its snowy head upward. Wings flapping hard, it rose into the air, yellow razor-tipped talons thrown out behind, a fat trout caught in their grasp.

Twenty feet up, the eagle lurched, as if righting itself from a phantom gust of wind. The trout slipped from its grasp and fell to the lake with a plop. The eagle let out a rhythmic, high-pitched chirping noise. The sound was sad and pathetic, not at all what I expected from such a magnificent creature. It flew close around our little boat twice, continuing its lament, before climbing high into the sky to resume its hunt elsewhere.

It wasn't until the eagle had flown out of sight that I realized I'd lost my fish too. I wasn't sure if the eagle's fish and mine had been one and the same, or if my fish had just taken advantage of the distraction to make its escape. Either way, the sighting of the eagle had more than made up for the loss.

"I told you this was a lucky place," said Dad, clapping me on the back. "But I think the fish have probably had enough excitement for now. Let's head in for lunch."

We motored back across the lake with the wind and sun and spray adding to our exhilaration. Back at camp, we found Mary dozing outside the tent. While cleaning the fish, we told her all about our morning. She nodded, smiled, and said she was glad we all had so much fun together.

Mary already had a small fire going, so we smothered the fish in lemon-pepper, butter and salt, wrapped them in aluminum foil, and placed them over the hot coals. While the fish were cooking, Kendra asked Mary if she wanted to come out on the lake with us after lunch.

Mary glanced from me to my dad and then shook her head. I pretended not to notice and gazed at the wisps of smoke drifting up from our fire. My dad was sitting next to me, with Mary across from us, and Kendra in between. Mary had set the chairs up the night before. She'd made the choice to be as far away from my dad as possible, so he and I could have our space and time together. As much as I hated to admit it, Mary was a sweet lady. I liked her. Even as I thought it, I felt I was betraying my mother, but then I let it go. The adults had made their choices, now I was making mine.

"Come on, Mary," I said, squinting up at her. "Please come with us. It's fun out on the lake. We might even see the eagle again."

She and my father exchanged a look and my dad patted me on the back.

"Okay, well since you've both invited me, I will. Thank you, Jason. Thank you, Kendra."

As we ate, I considered my father and the way he was, the life he led, and something that would have a profound influence over me clicked into place. Life could be wild and exciting and full of endless possibilities, or it could be safe, comfortable and boring, all depending on the choices one made. It was that simple.

I sure as hell didn't want to be the one snoozing in a comfy chair, waiting to hear about the adventures others were having. I was too much like my dad to settle for that. I decided then and there that a vicarious life would never do for me. I wanted more, no matter what the cost.

If only I had made that decision a year later.

Ruin

Upon returning to Arkansas at the end of summer, I felt lost. Most of our time in Wyoming had been spent in the wilderness. Fishing, camping, climbing, hiking - all the things I loved. Now it was back to school, which I hated, and dreary evenings watching garbage on the idiot box. Even the joy of seeing my mother again was short-lived. She was too busy with work to spend much time with Kendra and me. I was bored and miserable.

Then I heard about *the ruin* and everything changed.

*

"It's where the witch died," said Mikey, as I arrived at the cafeteria table. My three best friends were having lunch and talking. It was a dismal day in the middle of October. Mikey had already finished his food. Chris and Gary hadn't started theirs.

"I thought her place was out by Sunset Ridge?" said Chris, leaning forward on his long, lean arms.

"What witch?" I asked.

Mikey peered down his nose at me. We were the same age, but he was a lot larger. Big-boned and fleshy, with tightly cropped blond hair. These days his build would be considered normal, but back then he was our grade's fat kid. "You don't know about the witch, pipsqueak? You live under a rock or something?"

"The witch was a woman who killed and ate her husband," explained Gary. "Everyone knows about her. Her house is Witch's Hollow, on Sandown Road. People say she still wanders around there on full moons. It's a spooky-ass place, but nowhere near as bad as *the ruin*. Her husband is the *black cloak ghost* who hangs out near Curia Creek where she was supposed to have killed him."

"That guy's not her husband. That's the Indian chief," interrupted Mikey. "He haunts the creek looking for white boys to kill, because he's angry the townspeople killed the witch in his cave and built an insane asylum on top of it."

"It wasn't an asylum. It was a hotel!" insisted Chris.

"You're both wrong. It was a school for crazy kids. Kind of like this one." Gary crossed his eyes and let his tongue loll out the side of his mouth. We laughed, and he continued, "Listen, my grandpa told me all about it over the summer. If you want to hear the truth, here it is. Years ago when the witch was alive, two kids went missing. Their parents went crazy trying to find them. Everyone knew the witch did it, but they couldn't prove it because there weren't any bodies. Anyway, the witch's husband – who was also an Indian chief before she put a spell on him and made him marry her – told one of the boys' father where she'd taken the kids."

"The cave beneath *the ruin,*" said Mikey.

"That's right. Now the sheriff and his deputies went into the cave and came out a little while later with two body bags. No one knows if the boys who went missing were inside them bags because after they'd loaded them up they went back inside and came out with two more, then two more. They kept going like that for nearly an hour, sweating and hauling, until in the end there were thirteen body bags. Only a few of them were human shaped. The rest were either weird square things or big lumpy messes. But all of them were bodies."

"Bullshit," said Chris.

Gary gave him a long hard look. "You calling my Grandpa a liar, Chris?"

Chris shrugged. "Could be you, could be him. All I know is what I smell."

"Just tell Jason what happened next," said Mikey.

Gary took one of Chris's fries and munched it, mouth open. He swallowed it and belched in Chris's face. "Now what do you smell, wise guy?" he asked, before turning back to me. "Okay, so the townspeople joined up with the sheriff's men and they all went to Witch's Hollow to get the witch."

"Why would a witch live in a place called Witch's Hollow?" asked Chris, shaking his head. "I mean, if you were Dracula you wouldn't buy a house called Vampire's Rest, would you? Surely witches and vampires don't want everyone knowing where they live."

"Will you shut your smart-ass mouth and let him finish the story?" said Mikey.

Chris shrugged again.

"Thing is, the witch wasn't home," said Gary. "She was over at Curia Creek killing and eating her husband for telling tales on her. She knew the townsfolk were onto her, so after she ate her old man she decided to hide out in the cave. When the cops killed her, the old chief kicked her spirit out of his cave and she went home. Now she can be found either in her house or wandering up and down the road outside, in a state of eternal torment."

"Back up a second, wasn't the chief her husband?" I asked. "The one who haunts Curia Creek? How was he in the cave?"

"Different chief," said Mikey. "There were tons of them in those days. Tell them what she said before they killed her, Gary."

"Well, they finally found the old hag, half-starved and full-crazy, in the cave beneath *the ruin*. She was upset about killing her husband – because she loved him, I guess – but not about killing the kids. She said there was a soul-eating demon living in the depths of the river that flows through the cave and she fed it all those kids to keep it happy. Kind of like a public service.

"When the cops heard her story, they had no choice but to kill her on the spot and throw her body into the river. They said she could be the soul-eater's last goddamn meal. Then they built *the ruin* over the cave's entrance – except it wasn't a ruin back then, of course – and they used magic symbols and stones and arrow heads and Indian bones to keep the soul-eater from escaping."

"Horseshit," said Chris.

"You think so?" said Mikey. "My brother's been there and he told me all about the crystals and bones in the walls. He saw them with his own eyes. Why would someone go to all that trouble if there's nothing down there?"

"Where is it?" I asked. My guts were squirming with terror and excitement. "We should go check it out!"

"You're crazy," said Mikey.

"And you're a pussy," said Chris. "I'll go. I'm not afraid of Gary's grandpa's stupid stories."

Gary glared at Chris. "Okay, let's go. It's about our time anyway. All the older kids have been there." He turned to Mikey. "You afraid you're gonna come out in a body bag, fatso?"

"Yeah, one of the big lumpy ones," added Chris.

Chris, Gary and I laughed. Mikey took a fistful of Chris's fries and shoved them all in his mouth. "Uh-huh, that's right," he said, through a spray of mashed-up potato. "And you'll be in a square one, you pansy."

"Funny," said Chris. "Now will you two quit eating my food? I don't want to end up like Jason."

I didn't say anything. Chris was already as thin as me, but he was also tall, and athletic. Gary was all wiry red hair and Scottish brawn. Mikey was chubby, but strong as an ox. I was small and scrawny and I wore glasses; the nerd of the nerds. Deep down, however, I suspected I was braver than all of them put together. And when we got to *the ruin* I intended to prove it.

*

The rain continued over the next few days, and so did the stories about *the ruin* and the mysterious cave over which it had been built. In one of the tales, the cave was haunted by a slave who had been tortured to death for running away from his masters. In another, it was the site of a great massacre of Native Americans. The perpetrators changed with each telling. Sometimes it was other Natives. Sometimes white settlers. Sometimes a gang of runaway slaves.

The story I liked best was one about a race of humanoid lizard-people killing a group of explorers trying to find the source of the subterranean river that runs through the cave. According to this tale, the explorers had discovered a network of tunnels upstream from the cave. One of the tunnels led to an underground city inhabited by benevolent blue-skinned giants.

The blue-skins showed the explorers all sorts of advanced technology and explained how the Cherokee people had driven them underground many years before. Apparently, they had hoped to find peace in their hidden paradise but had instead met with constant attack from lizard-men and brutish Neanderthals - both of whom believed the world beneath ours belonged to them.

When the explorers left the protection of the blue-skins, the lizard-men ambushed them in the cave beneath *the ruin*. The reptilians slaughtered all but one of the explorers. After telling of his ordeal and the miracles he'd seen below ground, the lone survivor was locked up in the insane asylum that had been constructed above the cave during the expedition's absence.

While the rain continued to pour, and the stories grew ever more fantastic, I became more and more fascinated. Real adventure was beckoning. And it was only a short bike ride from my house. With any luck, I could rescue the besieged race of blue-skinned giants, defeat the

Falling to Fly

marauding reptoids, face down the resident soul-eating ghost, and still be home in time for dinner.

*

The rain stopped sometime Saturday night. Sunday morning was damp, cloudy and chilly, but dry. Shivering in my parka, I set off on my bike for Mikey's house. Gary and Mikey were neighbors, and I found them watching cartoons together in Mikey's living room.

"I'm not going anywhere," said Mikey, when I explained why I was there. "It's too cold."

"But it's not raining," I countered.

Mikey glanced at the window and gave a start. "Jesus Christ, what the hell is that?" he said, eyes wide in mock horror.

Chris was shuffling up the path to Mikey's front door, wearing a padded green jacket, full-face black woolen balaclava and yellow tinted ski-goggles.

"Chris disguised as a member of the I.R.A. by the looks of it," said Gary.

Mikey and I exchanged a look. "Must be a British thing," observed Mikey with a shrug. I nodded. Gary's parents were from Scotland and he often made references to things we didn't get.

When I opened the door, Chris pushed up his goggles and pulled the eyehole opening of the balaclava down over his chin. His nose was as red as Rudolf the Reindeer's.

"Perfect! You're all here," said Chris. "Are we going to check out *the ruin*, or are you pussies too busy snuggled up indoors watching cartoons?"

"I don't see how watching cartoons in the warm makes us pussies," said Mikey, with a shiver. "Come in and shut the door, will you?"

Chris raised an eyebrow. "You don't?"

Gary stood and walked over to the coat rack. He pulled on his red plaid jacket and wrapped his scarf around his neck, then threw Mikey's enormous brown duffel coat at him.

"Let's go," he said. "We'll get no peace from these two until we do. Never mind the weather or the fricking *Thunder Cats*. It's time, Mikey."

Half an hour later our bikes were leaning against the brambles at the bottom of the high gray wall surrounding *the ruin*. The four of us had climbed onto a narrow stone ledge and were peering over the wall.

This was my first glimpse of the place, and I was not disappointed.

An unkempt courtyard stretched between us and the main building. The flagstones were cracked and displaced. Clumps of grass and thorny shrubs pushed up through the gaps. The slimy, moss-covered stones, dew-spattered spiders' webs and sickly-sweet stink of decaying vegetation made the dank courtyard one of the spookiest places I'd ever seen. But it was nothing compared to what loomed behind it.

The ruin was wide and squat, two and a half stories high, with a sagging gable roof. The walls were made of thick blocks of gun-metal gray stone. Strange shapes and symbols, picked out in ivoried bone and glittering green and purple crystals, were set between the stones. Lines of flint arrowheads tracked along the mortar, pointing to tiny geode caves, where cracked frosted fangs surrounded mouths of dark quartz and amethyst.

The woodwork around the boarded-up windows and doors was warped, and the forest green paint on the frames and makeshift shutters bubbled and flaked. A pair of heavy iron handles protruded like broken bones from the cracked wood of the doors. They were bound by a thick rusty chain and sealed with an ancient padlock.

Chris nodded at the doorway. "Looks like they're pretty serious about keeping people out."

"Maybe. More likely they're trying to keep something inside from escaping," said Gary, making the sign of the cross. "I don't know about you guys, but I think I've seen enough."

I thought Chris was going to call him a pussy again, but he didn't. Instead, he slunk down off the ledge and picked up his bike.

"What are you doing, Chris?" I asked.

He shrugged. "We've seen it. That's why we came. There's no way I'm going in there. What if someone comes?"

"Are you serious?" said Mikey. "You dragged me away from cookies and cartoons to peek over a wall and that's it? What are you afraid of? You really think there's a soul-eater in there?"

"I thought we were all going in," I added, inwardly cursing myself for sounding so lame and whiny.

"Chris's right," said Gary, hopping off the wall to join him. "There's no way in. And you'd have to be crazy to break into a creepy stink-hole like that anyway. Let's go do something else."

"No way. I'll find a way in," I said, raising my chin and straightening my glasses.

Chris and Gary looked me up and down and laughed. "Sure you will, pipsqueak," said Chris. "In the meantime, we'll be at the park, listening out for your terrified screams."

Still laughing, they rode away. I was so furious and focused on proving Gary and Chris wrong, I almost forgot that Mikey was still standing beside me.

"They're just scared," he said. "And full of shit. They both know my brother told me how to get in there. They just don't want to go." His pale, doughy face crinkled into a frown. "Can't blame them for that, though."

"You know how to get in?" I asked.

Mikey nodded, making his double-chin quiver. "Sure I do. But I'm scared, too. Do you really think we should go in there?"

"I'm not asking you to come with me," I said.

"I know you're not, but I can't let you go in alone."

Mikey was a good friend. The kind you'd want to keep around your entire life. Unfortunately for me, within a few months I would forget all about him.

*

The way in turned out to be through a low window. One of the boards covering it was loose and could be pried open, according to Mikey. Once he pointed it out to me, I pulled myself up over the wall and dropped onto the flagstones. They shifted under my weight, oozing dark mud up over the edges. Mikey scrabbled over the wall behind me. He dangled from his fingertips, face up against the bricks, his feet about eighteen inches from the ground.

"How far is it?" he asked, without looking around.

"About a foot. You're nearly there. Just let go."

He did. The flagstone he landed on broke on impact, sending a splash of foul-smelling water up the front of my jeans. His legs trembled like Jell-O, and he staggered back and forth until he managed to slap his hands against the wall to steady himself. I tapped his shoulder and he flinched.

"You okay? You ready to check out that window?"

Mikey nodded and turned around. It wasn't until we were facing each other that he opened his eyes. There was a fine sheen of sweat on his forehead and the vein at his temple pulsed.

We weaved our way across the courtyard, dodging the dense shrubs and glistening spiders' webs, the stones beneath our feet making squelching and sucking noises. I glanced back now and then to make sure Mikey was still there. He followed me step for step, his eyes intent on a point somewhere between my shoulder blades. When we reached the window, he looked startled; as if he couldn't quite believe we'd made it that far.

"The big one on the left." His voice was a throaty whisper, and he was staring dead ahead, petrified. "Get your fingers under the bottom and pull it up."

The board was mulchy and eased back over its nails just as Mikey said it would. Once the nail-ends were cleared, it pivoted on the single nail that held it at the top. I pushed it round and up until it jammed firmly over the board next to it. There was now a dark square hole where the board had been. Big enough for Mikey to squeeze through, and plenty big enough for me.

I was about to climb in when Mikey put his hand on my shoulder.

"I want to go first," he said.

"Why?"

"Because if I don't I'll probably run away the moment you're inside."

We both laughed, breaking some of the tension we'd felt since entering the courtyard.

"Okay. Seeing as you put it like that. You want a hand?"

Mikey shook his head. He frowned at the black square, raised his flashlight, turned it on, and poked his head through.

"See you on the other side," he said, heaving himself over the sill.

The top half of his body disappeared, leaving his legs waving comically in the air for a couple of seconds before they slid from view as well. A series of thuds and thumps, followed by a curse, and Mikey's face reappeared at the window. "The floor's lower inside," he explained. "Come on Jason. I don't want to turn around until you're in here too."

He stepped back to make room while I gripped the inside of the window frame and pulled myself up. Folding my legs under and between my arms, I was able to enter feet first. Once my legs were through, I twisted my body around so I could lower them to the ground inside. My toes tapped against spongy plaster and I took one last look back across the courtyard.

A flash of yellow and orange caught my eye; Chris's ski goggles and Gary's hair. The expressions on their faces were an identical mix of awe and panic. I smiled and waved, then shimmied back through the gaping hole.

As soon as my feet touched the floor, I sensed something wasn't right. My neck hairs tingled and my stomach did somersaults. There was somebody in there, watching us. I was sure of it.

"We need to get out of here," whispered Mikey. "Something's...."

"I know. And we will. But we have to stay a few minutes at least. Gary and Chris are outside."

"So?"

"If we leave right away we'll never hear the end of it. But if we stay... well, they'll never call us pussies again."

"Uh-huh," said Mikey, staring at the floor. "But Jason, I don't care about that. I think someone else is in here."

I looked around. We were standing in a narrow corridor. A shaft of dingy daylight clogged with swirling dust motes came in through the hole we'd made. To both sides, further slivers of light pushed through the gaps in the boards over the windows. I could see maybe twenty or thirty feet along the gloomy hall on either side of us. There were wooden doors set at regular intervals in the wall opposite the windows. All of them closed. If there was someone in there, they weren't as close by as they felt.

The air inside *the ruin* was heavy and charged, as if a thunderstorm were about to break. It was also dead silent. The rumble of traffic from the main road had been audible from the courtyard, along with country music playing on a radio, and an occasional barking dog. But inside, even right by the window, there was nothing. All sound stopped where the darkness began.

Even though I'd always laughed at those dumb-asses in horror movies when they did the exact same thing, I just couldn't help myself: I shined my flashlight down the corridor and started to walk.

"Where are you going?"

"I don't know. You coming?"

Mikey made a whimpering noise and then shuffled up behind me. His flashlight cut a beam through the dark alongside mine. Both were trembling. We crept along the corridor, huddled close, studying the shadows and listening for any sign of movement.

Mikey gripped my arm and pointed to the door ahead of us. It was ajar. My heart pounded. As much as I wanted to turn and run, I couldn't. Something inside was telling me this was it; nothing would ever be more exciting than this moment, and if I backed off now, I would always regret it. I stepped forward and tried to push the door open, but it was old and heavy and wouldn't budge.

"Give me a hand," I whispered to Mikey.

Dutifully, he braced his shoulder into the door and drove forward. It scraped open, releasing an acrid, earthy stench that reminded me of mushrooms. I raised my flashlight and took a look inside. Old chunky furniture. A stack of books by a bedside lamp. Everything was covered in dust and cobwebs and cast long dense shadows. We were still standing on the threshold when we heard a creaking from down the hall. The sound resolved itself in a dull thud. My skin prickled and I almost lost control of my bowels.

Mikey pulled me back into the hallway. "Let's go!" he hissed.

I nodded and we dashed back in the direction we'd come. The beams of our flashlights jerked over the doors and shuttered windows, searching for the tell-tale square of light that marked the hole where we'd come in. We didn't find it. Instead, the corridor ended and we spilled out into a large hall of some kind. I dragged Mikey to one side where we fell against a wall. In mute agreement, we both turned off our flashlights.

I opened and closed my eyes several times, trying to get them accustomed to the dark. But it was impossible. If anything, it felt darker when my eyes were open than when they were closed. All I could hear was Mikey's ragged breathing and my heart thumping in my ears.

"Shh," I said. "Let's stay here for a minute. See if anyone comes."

Mikey swallowed hard, but said nothing. I counted in silence, to three-hundred. There were no more noises, but I still felt as though someone, or something, was watching us. I tried to push the feeling aside, to be rational, but it wouldn't go. It stayed stubbornly at the base of my skull, sending shivers of dread trickling down my spine. I needed to move, to do something to take my mind off what I hoped was my over-active imagination.

"You ready?" I asked.

"For what?" Mikey's voice was thick and croaky.

I shrugged. Then, realizing Mikey couldn't see me, added, "To look for the cave, I guess. It's why we came in here."

"Are you nuts? S-s-someone's in here," he stuttered. "And they're after us!"

"Calm down. Nobody's after us. We haven't heard anything since that door slammed. I bet it was just some other kids."

"You really think it was kids?" Mikey asked.

"Who else?" I held my flashlight under my chin and turned it on, casting my face in sinister shadow. "Was it the witch come back from the dead to hunt a nice juicy fat boy for her supper?" I cackled.

"Quit it," said Mikey, knocking the flashlight aside.

"I'm joking. Seriously, Mikey, there's nobody here. We would have heard them by now if there was."

We climbed to our feet and continued, keeping to the edge of the room like fishermen afraid to lose sight of shore. I figured we must be in some kind of dining hall. I couldn't see any tables, but it was too big to be anything else.

The interior wall we followed was made up of the same mix of stone blocks, arrowheads, bones and crystals we'd seen outside. My shirtsleeve and hand scuffed and scratched against them as we walked. I kept thinking there was something out in the center of the room, and every now and then I flicked my flashlight away from the wall, trying to catch sight of whatever it was. If anything *was* there, I never saw it. The room was too vast and dark; an ocean I dared not cross.

At one point, the feeling grew so strong I turned and shuffled sideways, my back to the wall, fingertips to rough stone, while the feeble beam of my flashlight swept back and forth, scanning the room. Mikey followed suit. I glanced at his light. It was weaker than mine, the beam a dim yellow.

I was so mesmerized by our synchronized sweeping arcs of bright white and dull gold that when the wall behind me abruptly disappeared, I stumbled and almost fell backward into the void. In my panic, I dropped my flashlight. It clattered to the ground and died.

"What happened?" Mikey swung his light around and shined it into my face.

"Stop that, will you?" I said, holding up my hand and turning away from him. A wave of nausea broke over me and I closed my eyes and waited for the black spots to clear from my vision. When they did, I looked down to see Mikey hunched over the broken remains of my flashlight. The lens had shattered, and he was flicking at the lifeless bulb. It made a sound like a tiny bell each time his finger struck it.

I knelt beside him and punched the floor in frustration. How could

I have been so clumsy? I snatched Mikey's flashlight and shined it at the wall. Or rather where the wall should have been before it had vanished. A high arched gap was cut into the rock there. I moved forward and felt a damp breeze upon my face. A stairway of rough-hewn steps led downward in a tight curve. In that moment, all fear and frustration melted away, replaced by the exhilaration of discovery.

"I think we've found it, Mikey. Look!"

Mikey wasn't as thrilled as me, but he let me keep his flashlight and followed when I started down the steps. The passage was tight, and seemed to grow narrower the further down we went. The sense of unease I had experienced since entering *the ruin* now increased as we wound our way deeper into the earth. At each turn, I expected to meet something fierce and deadly coming the other way. Sweat ran into my eyes. The breeze had died and the air was now clammy and uncomfortably warm.

Eventually the stairway straightened out and ended in a crumble of stones and pale sand. The river was there, lapping at a subterranean beach. My stomach was heaving and my heart skipping. It was difficult to breathe without gasping. The watchful presence was stronger in the cave than anywhere else. I stepped down onto the sand and made a move toward the swishing water.

"I've got to go. I can't stay here," Mikey whimpered.

"But the river," I insisted. "We should at least see the river."

I kept walking, but with every step I took closer to the water, the flashlight beam weakened and receded. A loud plop and swish echoed through the cavern and Mikey yelped like a kicked dog. Ripples spread out across the black surface of the river, catching the flashlight's feeble light. I saw them coming toward me, though I could see little else.

"Okay," I said, and turned around.

I went to step past him, but he blocked my way.

"Give me the flashlight. I'll lead us back up."

He sounded so desperate and afraid, I handed it over without argument, and fell in behind as he trudged up the stairs, slower than molasses.

I was worried about his dying batteries and kept telling him to hurry. It didn't make any difference; he just kept his own pace. In spite of being overweight, I knew Mikey was fit enough to make the climb with ease. Fear alone was holding him back. The same fear that made me push at his heels. We both felt it. Around the corner. Ahead. Behind. *Something* was there. Watching. Waiting.

When we reached the top of the stairs, I squeezed past Mikey and out into the wide room. My foot scuffed against something that skittered across the floor. I held my breath and Mikey let out a nervous giggle.

"It's only your flashlight, dummy," he said. "Come on!"

We hustled back along the wall to the corridor. As soon as we turned the corner, Mikey's flashlight died. He hit it against the palm of his hand a couple times but couldn't beat any life back into it.

"Batteries gone," he said, stating the obvious.

I smiled at him. Thin shards of daylight coming through the gaps in the window boards cast his face in a grayish glow. He looked pale, petrified and pathetic. But I'd never been happier to *see* anyone.

"Doesn't matter." I nodded at the corridor. "It got us here. Imagine if it had gone out back there."

"I'd rather not," said Mikey, trying to return my smile.

The sun must have gone behind a cloud outside, because inside *the ruin* it grew darker. The blackness of the large hall seemed to stretch over us, and we heard muffled mumbling echoing across from the other side. For the second time that day, we took off at a run down the corridor. Halfway down, the sun returned and a thick beam of light appeared from nowhere in front of us. We stopped dead, and I half expected tiny angels to appear in the golden ray, playing a fanfare on harps and trumpets.

The moment passed, and the sun slipped away again, leaving a cold steely shaft in place of the warm gold. Not that we cared either way. It was the opening where the light came in that mattered.

We wasted no time getting out. Mikey went first. I followed close behind. Once through, I found Mikey seated against the wall, breathing heavily and grinning from ear to ear. He didn't seem to notice or care that he'd pushed down another flagstone and was now soaking up the creeping black water with the seat of his pants.

The daylight was so bright I spent a few seconds blinking and shading my eyes until I became re-accustomed to the light. Eventually, Mikey stood and held up his palm. We high-fived, and as we did, we heard a loud whistle, followed by the sound of applause. Gary and Chris were sitting on the far wall with Mikey's brother and a few of the other older kids. They were all clapping and cheering. Mikey bowed, and we swaggered back across the courtyard like conquering heroes.

Somebody shouted, "Go Fatman and Ribbon!" Everyone busted up laughing. Including Mikey and me.

*

Over the next few weeks we became minor celebrities at school. We were the youngest kids who'd ever gone into *the ruin*. This meant we'd earned everyone's respect. We went from being two boys destined for obscure nerdhood, to Fatman and Ribbon, the fearless, if still pretty nerdy, young explorers. We both reveled in the attention.

What was more important to me, though, was our time in *the ruin* itself. I felt transformed by it. It sparked my imagination; filling my waking thoughts and providing a vivid backdrop to my dreams. At some point, I became convinced there was treasure hidden in there, and that I was the one destined to discover it.

"You can't go in by yourself," said Mikey, after I'd told him and the others about my plan. Me, Mikey, Chris and Gary were back in Mikey's living room, munching our way through a pile of his mom's world-class peanut-butter cookies. "And I'm not going back in, especially not if these jokers are gonna be hanging around outside like last time."

"Why not?" asked Chris. "You don't like applause? I thought we did a good job of keeping watch while you two scared yourselves to death."

"At least we had enough balls to go in." Mikey leaned forward and dunked his cookie in Chris's milk. "You think we don't know it was one of you who closed the board when we were inside? There's no way we'd want to give you two scaredy-cats the chance to mess with us like that again."

Gary and Chris exchanged a baffled glance.

"What are you talking about?" Gary's eyebrows knitted together, like they did when he was asked a question in math class, or before he was about to hit someone. "Have you gone soft in your head, Mikey?"

"The board was wedged up all the time you were in there," added Chris.

Mikey gave me a meaningful look. We'd all been friends long enough to know they weren't joking.

"Maybe we just missed it," I suggested.

Mikey snorted. "Impossible. I was desperate to find it. So were you. Both our flashlights were all over those window boards. We couldn't have missed it!"

"If you didn't see the opening, how do you think you got out?" asked Chris, perplexed.

"We're talking about the first time we went by," I explained. "When we came back later, it was open again."

"After the cave," said Mikey. "We should never have gone that far."

Gary's frown deepened. "Why? What happened?"

Mikey ignored him, and instead looked straight at me. "You felt it, Jason. I know you did. And you heard the door slam, and the voices in the hall. Now these guys say the shutter was up all the time. It's like we were being pushed and pulled around until we found that stupid cave! You might feel like you need to go back and search for treasure, but I don't. It's too damn creepy. Once is enough. If you're set on going, you'll be going alone. That's all there is to it."

In spite of everything, I was still set on going. My excitement and curiosity, not to mention my need to prove myself, were a lot stronger than my fear. Gary and Chris were avoiding my eyes. I knew neither of them would ever come with me. Mikey was right; I would be going in alone.

*

The first time I went back, I only made it ten or fifteen feet before the crunch of glass underfoot sent about twenty rats screeching away down the corridor, and me scurrying back to the courtyard.

After that, I resolved never to go there again. But three days later I was back. As terrified as the place made me, I couldn't get enough of it. I explored all the rooms I could get into, finding rags, beer bottles, candlesticks, dusty old books that disintegrated when I touched them, but no treasure.

Still, I kept going. Each time, I heard strange noises; whispering, creaking doors, slamming doors, shouts, screams. The sensation of being watched also got worse with every visit. But the worse it got, the more accustomed to it I became, until it was so familiar I no longer paid it any mind.

I made it back down to the cave several times, even going so far as to wade in the underground river, with its sinister swishing sounds and unexplained gurgles and plops. The malignant presence definitely felt strongest there. This I decided with the jaded detachment of a paranormal investigator. Every flashlight I ever brought down there seemed to falter the moment I stepped onto the sand, just as Mikey's had that first time.

The more I went to *the ruin*, the more my reputation as the crazy-brave kid grew. Not only was I the youngest person in our school to have ever entered *the ruin*, I was also the only one who went in alone, and stayed in there for hours on end. *The ruin* became my reason for living. It was a drug. I spent every waking moment either deep in *the ruin*'s rooms, corridors and caverns, or thinking about them. I was obsessed.

Then it stopped.

The presence had been growing angrier and more restless over the weeks, or perhaps I was just getting more sensitive and in tune to how it felt. I knew it despised me and resented my intrusion into its world. Still, I was drawn like a moth to a flame. Right up until the day the malicious spirit whispered directly in my ear. I didn't understand the words, but I felt the meaning well enough; it wanted death and revenge. For what, I didn't know, but it was enough to stop me going back ever again.

*

For a while, life went back to normal. I started hanging out with my friends and climbing trees again. I was happy at home, and content at school. My mom bought me a hand-held video game called *Donkey Kong*, which became my new obsession. Everything was going well, and just when I thought things couldn't get any better, my mom announced that we would be moving into a house. This was a big deal for us. We'd always lived in apartments, and the idea of Mom, Kendra and I having our own house was exciting beyond words.

The house was at the edge of town. There were wide sidewalks, grassy front yards, and neat two-story homes as far as the eye could see. The day we moved in I was in heaven. My new room was so big and airy I spent the afternoon zooming around it with arms outstretched like airplane wings. I could see trees and bushes and a neighbor's white-washed house from my window. It didn't look like anybody lived there. All the windows except the one at the top right opposite my room were shuttered closed.

That night we sat on the living room floor and ate take-out pizza, surrounded by boxes and good feelings. Mom made a toast to our new home, and the years of happiness we were going to spend there. We clinked our glasses of coke together and drank to it.

In the end, we only stayed there for three months.

*

When I came home from school the Monday after we moved in, I noticed someone in the window of the house next door.

"I think we have new neighbors," I told my mom later at dinner.

Mom frowned. "I don't think so, Jason. It's still closed up, isn't it?"

"Not all of it. One of the upstairs windows doesn't have any shutters. I saw someone moving around in there."

She gave me a searching look. "You sure? There's not supposed to be anyone there. As far as I know, the owner's not moving in until the summer. Keep an eye on it for me, will you? We don't want any vagrants squatting around here. Could be dangerous."

"Sure," I said, taking a bite of mashed squash.

After dinner, I stood at my window and watched the house next door for half an hour, feeling like a detective. I didn't see anybody, and began to wonder if what I'd seen earlier had been nothing more than the reflection of swaying tree branches on the windowpanes. Bored with my vigil, I went downstairs and watched TV for a couple hours.

I checked the room opposite again before going to bed, then lay on my side and stared out the window. We hadn't put curtains up in my room yet, and I could see the neighbor's roof and some tree branches. In the yellow glow of the streetlamps, it looked like an old sepia-tinted photograph. I was fidgety and agitated and couldn't sleep, so I got up and walked over to the window and looked across to the empty room.

This time someone *was* there. A tall, broad-shouldered Native American man with long white hair was standing up by the glass. His face was deeply lined and very pale. All sharp angles and aggrieved ferocity. He stared right at me, and a surge of emotion crackled across the space between us. Rage and hatred met incomprehension and terror. I tried to scream, but the sound choked off in my throat

The feeling was horribly familiar - from *the ruin*. I couldn't breathe or think. The demon, or whatever it was, held me with its gaze, refusing to let me go. Wave upon wave of resentment washed through me, suffocating my soul, spitting in my blood. Furious indignation bubbled inside me. I'd done nothing wrong. Nothing to deserve this.

Within my rage, I found the strength to pull away from the window. Tears streamed down my face. I threw myself onto my bed and pulled the covers up over my head, my skinny little body shaking in shock and fury and fear.

I thought the ordeal was over, but I couldn't have been more wrong.

All at once the stench of dead animals and rancid trash filled my nostrils. I gagged and squirmed further beneath the blankets. Something heavy slammed down on the bottom of my mattress and I twisted away from it. My knees folded to my chest, I gripped my blanket and held it tight above my head.

Whatever was there, moved. The weight shifted, and it began working its way up the bed, pressing down either side of me. Its breathing was hollow and ragged, and coming from somewhere over my stomach.

The ancient demon from across the street appeared in my mind's eye. I screamed for my mother, my sister, for anybody, to come save me. I knew the sound would be muffled, deadened by the sheets and blankets lying against my mouth. In a panic, I threw the covers forward, away from my face and over the head of my enormous assailant. The crumpled contours of the covers draped over the dome of his skull and the ridges of his shoulders and arms, giving the appearance of a vast rippling mountain range. I squirmed out from under him and ran to my bedroom door.

Flinging the door open, I pitched into the hall, screaming and gagging. Bile rose in my throat and into the back of my mouth. The light was too bright and I was disorientated. My sister's bedroom door swung inward. It was dark inside; Kendra wasn't there. I staggered away, and almost fell down the stairs. I gripped the bannister. The stairs, the walls, the door frames were all lurching in and out of focus. I screamed for my mom and stumbled down the first couple of steps, afraid I might fall. Halfway down, I crumpled to the floor and everything went black.

I don't know if I ever reached my mother. Not that night, nor any of the nights that followed. I do know these nighttime visitations continued for months, right up until we left the house and moved away. Every time it was the same; the same ghost-like ancient face, the same feeling of hatred, the same pursuit into my room and escape into the hallway followed by a blackout.

*

At some point this whole period in my life went dark, and I didn't remember it for nearly three decades. Not *the ruin*, not my friends, not the demon in the window - all of it was blocked out, leaving a year-long gap in my memories. Those memories were too terrible to revisit until I had no choice but to face them again many years later, when it was time to meet my childhood tormentor for the very last time.

Jason Pednault   Matheu DeSilva

Seeking Refuge

"What the hell is this?" said Bill, my mother's boyfriend.

At some point in the two years since we moved from Arkansas to Texas, my mom had started dating again. Unlike my father - whose relationship with Mary was stable and enduring - my mother was never really lucky in love. At that time, whenever she had a boyfriend they would usually last no more than a few weeks or months before disappearing off the radar, never to be spoken of again. Which was fine by me. As far as I was concerned, none of them were good enough for her. Mainly because none of them were my dad.

She'd met Bill at the mechanic's shop when she took our station wagon in for a service. He was big and scruffy, with grimy hands and thin, receding hair. His cologne was strong, but no matter how much he doused over himself, he could never get rid of the underlying stink of stale cigarettes and old engine oil.

I'm sure my mom would have dumped him after he'd fixed the leaky faucet and overhauled our apartment's asthmatic air-conditioners, but before she'd had the chance, our vacuum-cleaner had broken down as well. Now he had it in pieces on the kitchen floor, and was unwinding familiar-looking red, black and yellow coils from one of the moving parts.

"It's a scarlet king snake," I said. My missing friend must have slithered up the tube, and then been sucked into the workings when the machine was last turned on. "Or it used to be."

Bill squinted at me. "You tryin' to be smart boy? I know what it is. What's it doin' in yer mama's vacuum's what I'm tryin' to say?"

I shrugged and headed for my bedroom, too upset about the dead snake to entertain the idiot's question.

"Jason, you have to be more careful with those things," my mom called out, then softer, to Bill, she added, "I worry about that boy. He hardly speaks, and the only friend he has is this weird little girl he goes out critter hunting with."

"Probably jus' needs a daddy," said Bill, lighting a cigarette. "Someone to teach him man stuff and tan his hide when he needs it. Like right now, for instance."

"Don't even think about it. And don't you dare tell me how to raise my son. He has a daddy-"

"Yeah? Well why ain't he around to fix and service you and your things?"

"What did you just say to me?"

"You heard me, woman..."

Before the argument got into full swing, I closed my bedroom door and sunk to the floor. My animal index cards – a gift from my dad - were fanned out over the carpet. I studied them while trying to ignore the sound of my mom and Bill fighting in the kitchen. A dappled gray and green toad hopped over from under my bed. I picked it up and stroked its back.

A few minutes later I heard a knock at our apartment door. Mom and Bill continued to argue. I stayed hidden in my room. That left Kendra to shuffle down the hall to answer it.

"Jason! It's Ellen!"

I placed the toad on a cushion and headed out to the hall. Kendra glanced at the kitchen and rolled her eyes as she walked past me. Ellen stood in the doorway, grinning. With her wide mouth, large freckles and homely features, she and my pet toad could be distant cousins.

"Mom, I'm going bug hunting," I hollered.

Without waiting for a response, I ducked out the door and raced down the stairwell of our apartment block with Ellen. We continued to run all the way to the woods where we spent most of our time. Ellen was faster than me, but when she entered the trees I put on a burst of speed to catch up with her. As soon as I did, she swiveled around and blocked my path.

"You never told me about the rattler you brought home when you were five years old," she said, hands on hips. She was sweating and flushed from running. Strands of dirty blonde hair were plastered against the side of her face.

"So? Who told you about that?"

"No one told me. Your mom was screaming it to that guy when you came out of your room. Didn't you hear?"

I shook my head. "I wasn't paying attention."

She laughed. "He yelled back, asking what in the name of sweet baby Jesus she did about it. She told him she killed the darn thing with a broom. Is that true?"

I nodded.

"Ha! Your mom's a badass! Who was that jerk anyway?"

"Bill. Latest boyfriend."

"What a dick. He said she should have beaten you with the broom as well. Don't tell me you didn't hear that bit either?"

I shrugged.

"Who the hell thinks it's okay to hit a kid with a broom? Even if you did bring home a rattler! Why *did* you bring it home anyhow?"

I shrugged again. "I guess I didn't know what it was at the time."

"Bullshit! Ain't no snake you don't know, Jason. Speaking of which, I saw a big ole rat snake come through here not half an hour ago. It's huge! Wanna help me find it?"

I nodded, and we went into the woods. The snake was nowhere to be seen, but we did find some beetles and a few decent-sized spiders. We deposited them into the jars Ellen always carried in her explorer-bag (which I'd once mistakenly referred to as a purse, only to get a bloodied nose and a lecture from Ellen about her being 'no lady'). Once we'd secured the last perforated jar lid over a particularly feisty arachnid, she turned to me and frowned.

"So what's up?" she asked.

Even though we spent a lot of time together, Ellen and I didn't really talk much. Like most hunters, ours was a relationship of companionable silence and mutual reliance. As such, we tended toward more instinctive, telepathic forms of communication. Which is why she knew I had something on my mind. And why I knew she wouldn't let up until I told her what it was.

"I'm going to Alaska. My dad lives there now."

"I know. So what? You visit him every summer, don't you?"

"I'm not just going for the summer. I'm going to stay and go to school there in the fall."

"Was that your idea or theirs?" She nodded toward my apartment block.

"Mom thinks I need my dad. Dad thinks I need to come out of my shell more. I don't know. Maybe they're right. I just don't feel good here, you know? I don't really feel good anywhere."

"Thanks," said Ellen, giving me a dig in the ribs. She pried the hair from her cheeks and pushed it behind her sticky-out ears. "You know what I think? I reckon you'll love it up there. I would. You'll be catching moose, wolves and bears instead of snakes and spiders. Maybe you could bring me back a nice fur cloak with some sharp bloodstained fangs still hanging out the bottom when you come back to visit."

I laughed. "Maybe."

"Before you go, though, you best help me find this rat snake. If you don't I'll never forgive you." She glared at me, but the side of her mouth twitched in amusement. "Come on, I want to find that sucker before sundown."

We wandered through the woods, searching in silence, same as ever. But something was different between us. There was tension; I could tell Ellen was on the verge of saying something every time we stopped to check for signs of the snake. She didn't, though. Whatever was on her mind, she wasn't prepared to break our unspoken code to share it.

Just before dusk, the snake appeared on the trail in front of us as if by magic. Ellen hadn't lied. It was easily the biggest rat snake either of us had ever seen. We stood close together and watched it slink off the trail and over some sprawling tree roots. Neither of us made a move to catch it.

When it slipped out of view, Ellen clapped me on the back. "Sun's almost down, but not quite. You did it, Jason. I knew you would. I'm gonna miss your sixth sense for tracking critters." She turned away, rubbing her eyes like somebody waking from a deep sleep, and led us out of the woods.

*

Kendra and I flew into Anchorage on a fine clear afternoon at the end of July. I had received some fantasy books for my birthday, including *The Lord of the Rings,* and what I saw out the window gave me a thrill of recognition. Alaska was a landscape of raw adventure. Monstrous, craggy peaks, capped and streaked in snow, densely forested slopes, pale green pastures wrapped around lakes of turquoise and glittering sapphire. I imagined bears, and dragons, and quests, and treasure.

Anchorage itself wasn't so appealing. The right-angled grids of its streets and the banal uniformity of its high-rises seemed so alien and out of place between the rugged curving mountain range and the wide pristine bay – a cancer of rigid order chewing at the edge of wonder.

Our first two weeks in Alaska were spent camping and fishing, and it was everything I'd dreamed it would be; long days of sunshine, fragrant with wild flowers and pine resin, cool lakes, fat rainbow trout, twilight skies teeming with stars; it was heaven. And it was over all too quickly.

Kendra returned to Texas for a late-summer program she'd enrolled in, and Dad, Mary and I went to their new house out on the masticating edge of Anchorage. I was horrified when I saw where they lived. It was a duplex in a brand-new development. Concrete roads and sidewalks, with barely a tree in sight.

When we pulled up outside, Mary noticed the expression on my face and patted my hand. "Don't worry Jason, you'll soon settle in. We want you to feel at home and be happy here."

I wandered inside in a daze. Mary showed me my room while Dad brought up the bags. They were talking all the time, but I didn't hear a word. Once they left me in peace, I plucked *The Return of the King* from my bag and lost myself in its pages.

The next day I went out to explore the neighborhood. I found a small copse a mile or so from the house. The pine trees were too dense and stunted to offer much in the way of climbing, but at least it was green and quiet. I stayed there all morning.

When I arrived home, I found our front door had been left wide open. I rushed inside and stood in the entranceway, stunned. The first thing I noticed was the couch had been moved. It was also a different color and style. So was the carpet. In fact, everything had changed, right down to the pictures on the far wall.

My dad made his living as a flooring installer. I'd seen him work. He was fast and efficient. Even so, it would've been impossible to remove all our furniture, fit new carpets, and bring in all this new stuff during the three hours I'd been gone. I couldn't understand it. My puzzlement turned to panic when a man and woman I'd never seen before came through from the kitchen holding coffee cups.

"Who the hell are you?" said the man. "And what in God's name do you think you're doing in my house?"

I glanced back and forth between them, confused and frightened. He was angry. She appeared concerned. He was small and cruel looking. She was overweight and sweaty. Both wore identical green and purple plaid flannel shirts.

I mumbled an apology, backed out the door and ran to the sidewalk. The plaid couple followed me as far as the front porch, where they stood sipping their coffee, watching me. I stared back at them and tried to get my bearings. Mary's car was parked outside the next duplex along, and all at once I realized what had happened. I'd gone into the wrong house. I looked up and down the street. Every house was the same. The only difference were the cars in the driveway.

I shuffled away from the plaids' place and trudged up our driveway. Mary waved and smiled from the front window. I waved back, but couldn't return the smile. No matter how much we all wanted it, I knew I would never feel happy and at home there. If the wilderness of Alaska had felt like heaven, then the suburbs of Anchorage were definitely my idea of hell.

*

A few weeks later, I started junior high. Everyone had to switch schools at the beginning of seventh grade and I hoped I might be able to slip in unnoticed. Unfortunately, all the kids in my junior high came from the same elementary school. All the cliques were well established, and it was made clear from day one I wasn't welcome in any of them. I wasn't particularly bothered as I considered myself to be a natural loner (by this time I'd already forgotten Mikey, Chris and Gary), and once the bullies grew bored with my lack of reaction, I was pretty much ignored by everyone.

I'd planned on using my evenings to explore what nature there was around our neighborhood, but those plans were soon stymied by the weather. It was shocking how fast winter came on in Alaska. Each day grew colder and shorter than the last, and it wasn't long before I had to travel to and from school in total darkness, wrapped up like Scott of the Antarctic. Time outside was limited to hustling as fast as I could between buildings, or vehicles and buildings, seeking warmth and shelter as if my life depended on it. Which, of course, it did.

In that frozen land of winter, the school library soon became the most important place in the world to me. Not only was it a cozy haven where I could while away lunchtimes and recesses in peace, it also had the most extensive collection of fantasy and sci-fi books I'd ever seen.

They even had a range of choose-your-own-adventure books. Something I'd never seen before. In these books readers are given a choice at the end of each chapter that defines the path they take through the rest of the story. Each decision leads you to a different page, where you either make it to the next stage or die a gruesome death. It was much like the video games of today, but with a little more imagination required.

I would spend hours reading and re-reading every book, following the maze of chapters to experience every possible outcome. I thought if I studied them, I'd find the key to becoming a successful adventurer.

In reality, what I learned was once an adventure was undertaken, things would usually end badly, regardless of how well they were thought through. Success or failure had less to do with being prepared, brave, or clever, and more to do with the arbitrary whims of the gods.

*

It was mid-February before I made my first and only friend in Anchorage. A spell of unseasonable mild weather meant I was able to venture out to the nearby woods, so I woke early Saturday morning, ate some hot buttered toast, stuffed a book into the pocket of my parka, and set off.

After wandering the trails for a while, I found a snug nook beneath a large tree with a shelter of low branches and a carpet of soft dry pine needles. I settled in and began to read.

Sometime later I saw someone coming along the trail. It was a boy about my age, in a parka almost identical to mine. The only difference was the color; mine was dark navy-blue, his was a dirty orange.

The boy was bent over the trail, walking slowly. Every now and then he would prod at the ground with the toe of his boot or a gloved finger, turning over leaves or scrutinizing a displaced twig. He was tracking something, and when he got close to my tree, he stopped, straightened up, spun around three hundred and sixty degrees, then dropped into a crouch and looked under the branches directly at me.

"What you doing?" he asked. His eyes were pale blue and his washed-out face was covered in a constellation of dark freckles.

"Not much. You?"

The boy seemed curious and friendly, not at all like the hard-faced kids from school. Still, I was wary. Aside from my dad and Mary, I hadn't had a proper conversation with anyone in months.

"Hunting," he said. "You wanna come and watch?"

I thought of Ellen, and realized how much I missed her and our bug-hunting trips. I wanted so much to say yes to the boy, but something about his invitation offended me. I wasn't sure what it was until I re-played it over in my head. Ellen and I had been hunting *partners*. If anything, she'd deferred to me as the expert. This guy assumed I knew nothing. He'd asked if I wanted to *watch*.

"I don't need to watch. I know how to hunt," I said.

His eyes narrowed. Then he shrugged and broke into a wide smile. "Cool! You can help me then. I'm Daniel Roach. You?"

I told him my name, and he nodded, stood, and started walking away.

"Come on, Jason, let's go," he called back. "I want to find at least a couple more snare spots before we come back and set them this afternoon."

"Snares?" I asked, scrabbling out from under the tree.

"Yeah, snares. I thought you said you knew how to hunt. Or don't you use snares? I guess you only shoot moose and bears, huh?" His face was deadpan. I didn't know if he was making fun of me or not.

"No. No big game like that. I mainly hunt snakes. And spiders. Sometimes beetles and other bugs, too. But mostly snakes."

"Snakes, eh? Thought your accent was strange. Do you eat snakes where you come from then?"

"No, we're not stupid. Why would we eat them? Most of the ones I hunt are poisonous anyway." I was trying to impress him, but he wasn't fazed.

"That so, huh? Don't know why you'd want to hunt something you're not going to eat. But there's southerners for you." He stared at a gap between two young trees. "This way," he said, and took off along a side trail.

"So, what are we hunting?" I asked, after we'd been walking a while.

"Rabbits." He screwed up his nose, bared his top front teeth and made a rabbit noise by sucking at his bottom lip.

"I know what rabbits are."

He grinned. "That's something at least. You know how to set a snare to catch one?"

I shook my head.

"Don't worry. You will by the end of the day." He glanced at his watch. "Meet me back here at two o'clock, okay?"

I looked around. We were in the middle of the woods. "Here?"

"Uh-huh, right here." He scraped out an X in the dirt with his heel, then turned and ran off. "Catch you later, Jason," he shouted over his shoulder. "Don't forget. Two o'clock."

I went home for lunch, and then headed straight back to the woods.

I was a still a little way away from our meeting spot when I heard him calling my name.

"Hey Jason, I thought you weren't going to come. It's already quarter-to-two." God knows how he knew I was there. I couldn't see him. Even in his bright orange parka.

"That makes me early," I called back.

"And keen to learn, I hope," he returned.

I walked into the clearing and found Daniel standing in the middle of his X. In one hand, he held a ball of black twine and a few loops of wire, in the other a pair of pliers and a pruning saw.

Once I'd watched him set the first few snares, I asked if I could have a go. He eyed me skeptically but handed over the kit. I made the loop of wire and tied it to the string. Then I found a suitably springy sapling, cut it at an angle, the way I'd seen Daniel do it, and then tied the snare to it. Next up I wedged the sapling into the ground over a narrow passage between two rocks; a place Daniel assured me was a regular rabbit thoroughfare. Lastly, I put two small sticks into the earth either side of the snare, to hold it in place and stop it from spinning.

"Nice job," said Daniel, with a nod. "At this rate you'll soon be forsaking your serpent slaying in favor of the art of trapping the mighty rabbit."

I smiled. "I told you already; I don't slay them. I keep them. You really think we're going to catch some rabbits?"

He regarded me as if I might be a bit backward. "Nah, but just in case let's come back early in the morning and see, eh? I'll come by your house and get you. You only live around the corner from us." He squinted at his watch. "I better go. My dad will be home soon. Catch you later."

That evening, I asked Mary if she knew Daniel and his family. I wasn't surprised when she said she did. She already knew everyone in the neighborhood. She told me Daniel and his dad lived by themselves two blocks away. As far as she knew, he was an only child whose mother had run off not long after he was born. He was a year younger than me, and a sixth grader at the elementary school that fed my junior high.

Daniel knocked on our door at seven-thirty the next morning. The sky was black and starless, and it was still dark by the time we reached the woods. At the site of the first snare, we found a dismembered rabbit's foot caught in the wire. The rest of the rabbit was gone. A runnel of darkened earth marked the path of its escape. Daniel took off his gloves and loosened the snare. He went to hand me the small furry limb, but I recoiled away.

"What's the matter? You never heard of a lucky rabbit's foot?" he asked. "If you don't want it, I'll keep it. I could always do with more luck."

"It wasn't so lucky for its previous owner," I observed.

"You think?" Daniel stood and looked around. Then he shook the bloodied end of the paw at me. "I don't see this thing's previous owner anywhere near here. Which means he escaped. I'd call that lucky. Sure, his friends may take to calling him tripod, but that's a whole lot better than being in a stew pot then my belly." He slapped his stomach for emphasis.

"True. But man, that's pretty gross. It chewed its own leg off to escape!"

"Uh-huh, that there's called survival instincts." Daniel nodded sagely, depositing the foot into his coat pocket. "Let's hope they all didn't get the same idea. Otherwise I might have to start agreeing with you that rabbit paws ain't so lucky after all."

Of the five snares we set, three were empty, one yielded the paw and the last a whole fat rabbit.

Daniel whistled softly at the sight of it. "He's a big one. Almost makes up for the others coming up with nothing but a small scrap of luck between them. Speaking of which, you do know what we are witnessing right now is called beginner's luck, right?."

"Uh-huh," I said, mimicking the sound he made when humoring me.

The big rabbit, brown-gray and stiff with rigor mortis, was wedged between the two rocks where I'd placed my snare. Its eyes bulged in glassy accusation, and I found them difficult to look at. I'd killed plenty of bugs and spiders, and even a few snakes, but they'd all been accidental victims of neglect or clumsiness. I'd never killed anything more than a fish on purpose before, and I didn't know how to feel about it. I wanted to lean down and close its eyes like they do in the movies when somebody dies.

Before I could, Daniel was already kneeling beside it, loosening the snare and mumbling some kind of prayer in a language I didn't understand. A knife appeared in his hand and he gutted it right there on the trail. Once finished, he wiped smears of blood on his forehead and cheeks. Then he stood and painted my face the same way.

"We got native blood in my family, according to my dad, who's mostly full of crap. I mean, look at me! I'm paler than Crazy Horse's ghost and speckled in Irish freckles. Still, I like saying their words and doing the blood thing if I ever catch something. It's respectful. Now, you got to pick up the rabbit and hand it to me as part of the ceremony."

I did as he said.

"Cool. That means you gave me your first kill and I get to take it home." He grinned and I was about to protest when he held up his hand. "It also means we're brothers in the hunt, kind of like blood brothers. That's better than normal friends." He rummaged in his pocket and handed me the crusty bloodstained paw we found at the first snare. "That's yours."

This time I took it. We shook hands, sticky and slick with rabbit blood.

*

Daniel was a real outdoorsman and I learned a lot from him. He was also a joker with a wild imagination. At first, I couldn't understand why he didn't have more friends. But then I realized that like me, he was a loner at heart. He didn't want a whole bunch of friends any more than I did.

For me, being around a lot of people, especially if they're hyped or excited, always made me queasy and uneasy. I couldn't bear being at sporting events or amusement parks. Even school made me anxious. The only place I ever felt truly relaxed was out in nature. Either alone, or with people I knew and trusted. Daniel was different. He was fine around more or less anyone and thrived as the center of attention. He was a loner for another reason, and it wasn't long before I discovered what that reason was.

As winter set back in, Daniel and I would hang out at my place, or outside, but never at his. We watched cartoons or played chess or Dungeons and Dragons, and on blizzard-free days went out and played in the snow by my house.

One of our favorite things to do was rig up sleds with towing harnesses and let our dogs pull us around the neighborhood. We were both small for our ages, and the dogs never had a problem hauling us. My team consisted of our two Cocker Spaniels, a father and son pack named Chico and Dusty. Daniel had a lumbering Saint Bernard and Malamute mix called Griz who could never keep up with my two. Chico and Dusty would scamper across the snow with me gliding behind, while Griz trudged along dragging Daniel at half our pace. It was pretty clear Griz believed the whole Husky gig was beneath him.

Three-wheel ATVs were also really popular back then. Both our dads had one. My dad allowed us to use his in exchange for me doing chores around the house. We could only drive it around the neighborhood, but that was fine; we had nowhere else to go anyway. When there was a lot of snow we would clear a path for the ATV and pile all the snow we'd moved into a crude ramp off to one side. Daniel would put on his skis and I'd pull him down the street and launch him into the air off the ramp.

One time we had so much snow and wind, a natural berm had formed better than any ramp we'd ever made. My dad helped us scrape out the path for the three-wheeler and secure the towrope. Then he stood back to watch.

I looked back. Daniel gave me the thumbs up and I opened the throttle. My eyes were streaming as soon as I took off. The rope went taut behind me. I passed the berm, high up on my right. At the end of the track, I braked, skidding to a stop. The rope was already loose, discarded. Daniel dropped out of the air and landed some distance in front of me.

"Nice job," my dad called out from our front yard. He was clapping. "You must have been at least fifty feet in the air, Daniel! Don't you have any fear? Well rode as well, Jason. You both timed it to perfection."

Daniel and I exchanged satisfied grins.

"Again?" I asked.

"Definitely. I want to try to reach sixty feet this time. Seventy the next!"

We went up and down all afternoon, me riding the three-wheeler, Daniel throwing himself higher and higher into the sky. My dad had gone in after the first four or five runs. But he came to the window now and then to watch and wave, and shake his head in amazement.

When it was starting to get dark, we decided to make one more run before calling it a day. I set off down the track, but twenty feet short of the ramp the three-wheeler sputtered, stalled, caught, lurched, sputtered and lurched again. It was running out of gas.

I kept pressing the throttle, trying to squeeze the last dribble of fuel through the carburetor; the last drop of fun from the afternoon. The rope slackened, then tensed as a final burst of fuel propelled me forward. Daniel, off balance now, toppled over and smashed into the ramp. The rope was caught around his wrist, so I ended up dragging him through the berm and dumping him onto the hard, compacted snow of the track.

The three-wheeler's engine choked out and died, and I jumped off and ran to check on Daniel. He was slumped on the ground, arms over his head, still caught up in the rope. Before I got to him, someone grabbed me from behind and spun me around.

"What the hell you think you're doin', boy?"

Alcohol steeped spittle sprayed my face and I clamped my eyes shut. The man shook my shoulders so hard my head snapped back and forth. I took a peek at him, and saw a red face, twisted in rage. He was somewhere around my dad's age, and was wearing a filthy black and white baseball cap. Beneath the cap, he had a dark chevron mustache and pale bloodshot eyes. I'd never seen him before, but I knew who he was. The icy blue irises were the exact same color as his son's.

"Dad, leave him alone, please! I'm fine!"

"What you say to me, boy?" Daniel's father hurled me into the snow, then stalked toward his son.

My glasses had been knocked off, and I scrabbled to find them and put them back on again. When I did, I was relieved to see Daniel really was okay. His wrists were untangled now and he was standing in front of his dad, head bowed, as if praying.

Daniel's father was a big man. He was leaning forward and swaying from side to side. I couldn't hear what was being said, but I could see Daniel talking rapidly, his eyes still cast down. Out of nowhere his father's hand came back, formed a fist, and crashed down into the side of Daniel's skull. My friend collapsed as if he'd been shot. He was unconscious before he hit the ground.

"Hey Jimmy! What the hell you doing?" My dad was stalking across our front yard, eyes blazing. "The boys were only playing. Leave them alone."

"Mind your own damn business, Cliff. My Danny could've hurt himself bad playing that damn fool game."

"I was watching them. They were fine. Right up until you showed up and hit the poor boy that is."

"You come any closer and I'll beat you all up and down this street, you bastard. Your boy can watch me do it and then I'll do the same to him. My Danny could've hurt himself bad."

"Okay, Jimmy, you said that already." My dad kept coming until he and Jimmy Roach were toe to toe. "You don't want more trouble than you can handle, Jim. Now calm the hell down and let me help you with Danny. How about we bring him inside and see if he's okay?"

Jimmy Roach spat at the snow by my dad's feet. "Don't need your help. Don't need nobody's help. He's my boy and I'll take care of him."

While our fathers argued back and forth, Daniel woke up and climbed groggily to his feet. He staggered a little and shook his head. Then he winced and bent over, hands on his knees. He retched and vomited into the snow. Both men stopped to watch.

"See?" said Jimmy. "The boy's fine. Come on Danny. We're going home."

Jimmy Roach grabbed the hood of Daniel's orange parka and walked away, dragging his son behind. Daniel's legs slipped and stumbled, trying to gain purchase on the slippery ground. His pale eyes found mine. They were glazed and unfocused, and I'd swear he didn't even recognize me.

*

When I next saw Daniel, he didn't mention what had happened that afternoon with his dad. I never brought it up either. In the weeks that followed, he would often turn up at our house with bruises and other unexplained injuries. By unspoken agreement, neither of us ever mentioned them. Only my dad or Mary – who didn't know Daniel like I did – would ever say anything. When they did, Daniel would tell them some outlandish story to explain away his black eyes, swollen ears, the purple finger-shaped marks on his arms, torso and neck.

The tales he told were usually as funny as they were unlikely. Despite our concern, Daniel would soon have us laughing at his stories of foolhardy exploits and comedic clumsiness. Of course, we didn't believe a word he was saying. But that wasn't the point.

Daniel's cheerfulness and positivity in the face of all the abuse he endured was both inspiring and bewildering to me. My family never abused me. I was well loved and cared for. But still I always felt depressed, discontent, and I couldn't understand why Daniel didn't feel the same way. Truth be told, Daniel's attitude to his suffering unnerved me. It made me question what right I had to be unhappy.

I didn't *want* to feel the way I did. But being unsettled and jittery and lost was not something I felt able to change through willpower alone. Instead, I sought to escape into books and video games, with friends, or television. But try as I might, I could never master Daniel's skill of being at peace with the world no matter how bad things got. Not unless I was in the wilderness. Only in nature did I feel at peace and at home.

*

When I finished seventh grade, I moved back to my mom's. I don't remember whose decision it was, but I'm sure it had a lot to do with me not feeling any better in Alaska than I had in Texas. I was the same quiet, sad, awkward kid no matter where or with whom I lived.

It was good to see my mom and sister and the sunshine again. Ellen slotted back into my life, replacing Daniel as my only friend in town. Pretty soon, things returned to just how they'd been before I'd left.

Then Adrian arrived, and everything changed.

Jason Pednault   Matheu DeSilva

Presa River

Of all my mom's boyfriends, Adrian seemed the most harmless. He arrived on the scene halfway through my first run at eighth grade, though at the time I barely noticed this nervous forty-year-old engineer who couldn't believe his luck at finding his first real girlfriend.

He bumbled around our apartment, knocking things over, apologizing, and feigning interest in Kendra and me - in between trying to discreetly smooch with my mom. A consummate geek, the only vaguely disconcerting thing about him in the beginning was that he looked like a bigger version of me. He was rail thin, had unruly brown hair, a goofy smile, and big square glasses. Unlike me, however, he was clumsy to the point of slapstick.

Unfortunately, he was also a man with a plan.

Ever since returning from Alaska, I'd retreated further into my shell. I hardly spoke, preferring instead to escape inside a book or anesthetize my mind with video games. At school, I was so bored and disinterested I would literally doze away the days; sometimes I slept so deeply even the shrill bell that announced the end of class couldn't wake me.

The fact I was so quiet and unresponsive caused my mom to worry. I was on her mind a lot, and the more she concerned herself with me, the less attention she had left for Adrian. Having finally secured a girlfriend, I'm sure having to share her with a sullen teen like me irked him no end. Kendra was older, and a social butterfly who was rarely home. So the only thing standing in the way of his chances of an undisturbed and blissful honeymoon period with my mom, was me.

The first time I realized my klutzy would-be-step-father might not have my best interests at heart, was the day my mom received a call from my homeroom teacher. The school was worried about my poor grades and lack of interest in classes. Narcolepsy was also mentioned. As was the concern that 'problems at home' might be the root of my issues.

"Jason, you need to start taking school more seriously," she suggested, not unreasonably, while we were eating dinner.

I shrugged. "I try Mom, but it's so boring. I don't understand why they want to teach us half the stuff they do. And I don't really like the way they do it."

"It's not up to you to question the education system," said Adrian.

"It works for everyone else."

My mom shot him a look, and his hand quivered, making him dump a forkful of gravy-soaked chicken down his tie to splatter in the crotch of his beige slacks.

Kendra and I snickered while he tried to retrieve the fugitive chicken with a succession of paper napkins. "I'm sorry," he said, and I thought he was referring to the mess he was making until he added, "I don't want to interfere, but shouldn't we be looking for some kind of professional help? There are people and places out there who specialize in helping young people like Jason. Maybe we should speak to the school about assigning him a counselor. At the very least, it might help him to talk to someone neutral about his problems."

No one spoke for the next few minutes. My mom picked at her meal without taking a bite. Kendra scarfed hers down and then asked to be excused. Mom nodded and Kendra left the table. Adrian wet some napkins from the water jug and scrubbed at his pants. He cleaned up the gravy but left himself looking as if he'd pissed himself.

"So, what do you think, Jason?" said Mom, breaking the silence at last. "Do you want to talk to the school counselor about your grades?"

I shrugged. What harm could it do? The counselor only worked during school hours, which meant I would have to skip classes to see her. We would eat cookies, drink milk, and talk about me taking more interest in my studies. No problem. I could go through the motions if it made everyone feel as though they were doing something useful. Adrian smirked at me. I looked at his pants and smirked back.

*

The following Tuesday I had my first meeting with Miss Sanchez, the school counselor. I arrived early and waited in the deserted corridor for the clock on the wall to hit one o'clock, the time of our appointment, before knocking.

"One moment," she called out, before immediately opening the door. She tilted her head back and looked at me down the stubby length of her nose. "Ah, yes, Jason?" Even with her head at such an odd angle she still had four chins where her neck should have been.

I nodded.

"Good! Now how about you come right in so we can have ourselves a cozy little chat, sugar."

Falling to Fly

She ushered me into her office. The famous cookies and milk were on the desk, awaiting me. I sat down and stared at them. She squeezed into the pleather chair behind the desk and began swiveling from side to side. All the time eyeballing me from beneath her helmet of dyed blonde curls.

"You hungry, sugar?" she asked. Her tone startled me. It was more taunting now than friendly; not at all what I'd expected. When I'd seen her out in the hall she always seemed so nice. But this was the nineteen-eighties, and no one checked what happened behind closed doors in a public school. "What's the matter? Cat got your tongue? You want one of them cookies you're gawking at? Reach right out and take one. They're all yours, sugar." When I didn't move, she leaned across the desk, plucked a cookie from the plate and handed it to me. "Eat it!" she commanded, before collapsing back into her chair from the effort.

I took a bite, and she resumed swiveling. The cookie was stale, but I ate it anyway. The milk was fat-free.

As if reading my mind, she said, "I'm losing weight. That's why the milk is so insipid. Don't want any fresh cookies around either. So I hope you're going to eat them all up for me like a good little soldier."

She chuckled, and I frowned.

"So, now we're friends, maybe you can tell me what the heck's going on with you, Jason?"

"Ma'am?" I asked.

"Ma'am?" she repeated, in a squeaky voice. "You know what I'm talking about. Don't play dumb. Why do you think it's okay to sleep in class, huh? Why don't you show more respect to your teachers?"

"I do show them respect."

"Oh yeah? By wasting their time and disrupting their classes?" She made a series of piggy snorting noises while holding my gaze, then continued, "I don't think so, kiddo! Now what's your problem? Don't you want a future?"

I shrugged, not knowing what to say. Miss Sanchez snatched up her notebook and scribbled something in it. Then she leveled her eyes at me again, shiny black beads in a sea of blue eye shadow.

"I'm not going to leave off until you tell me what your problem is. Now you can start showing your enthusiasm for our little meetings by eating another one of them cookies. I don't want them on the table looking at me anymore than I want you wasting your life away and bringing shame on your poor mother."

I took another cookie and munched on it, hoping it would calm her down.

"Ah, so I touched a raw nerve, huh?"

"What do you mean?" I asked, through a mouthful of moldy cookie.

"When I mentioned your mama you should have seen the look on your face. Oh-ho! I've been doing this job a long time and I know when I've hit a bull's eye, sugar."

I shook my head. "My mom's fine. She's not the problem."

Her eyes narrowed. "Then it's your dad. What's the matter? Does he... you know, make you do things? Does he touch you?"

"What? No! My dad lives in Alaska-"

"Ah, so now it makes sense. You don't like that your parents split up and you can't handle them being with new people. Is that it?"

I thought of Adrian. "Sometimes." Then Mary. "No, that's not it. I don't know. I just feel like I don't fit in anywhere. And I get tired a lot." Tears welled in my eyes. I blinked them back and wiped my cheeks on my sleeve.

Miss Sanchez leaned forward and seized my hand, crumbling cookie and all, in her pudgy mitt. "Well you need to get over it, sugar," she soothed. "Otherwise your whole life's going to be nothing but a shit hill of pain."

I went to see this woman six times. Each time it was the same bizarre sequence of warm welcomes, personal attacks, wild assumptions and companionable cussing. The end result was a referral to a specialist in personality disorders. Incredibly, the referral was for me, not her.

*

The specialist was a guy named Jeff. We met after school at a pizza restaurant with a collection of video game consoles. Unlike Miss Sanchez, this guy was a real pro. He bought me pizza, we chatted for a while, and then he let me run to the back of the joint to play *Galaxian*. I thought he was a cool guy.

We had three half-hour sessions, each followed by an hour of video games while he sat in the booth, smoked and wrote his reports. He asked the same sort of questions as Miss Sanchez, but in a nicer way. Fascinated by my feelings about my parents' divorce, he seemed convinced it was the root cause of all my problems. Of course, I agreed with him. Why wouldn't I? He was picking up the tab for the pizza and gaming time. The least I could do was make him feel like he was doing his job well.

The fourth and final time I saw Jeff was at our apartment. School had finished for the year and I'd been out hunting scarabs and scorpions with Ellen. When I got home, Jeff was at the door saying goodbye to my mom and Adrian.

"Hey, Jason. Good to see you, buddy," he said, when he noticed me coming down the hall. "I've been having a talk with your parents, and we have some great news for you."

Parents? He knew Adrian wasn't my dad. I glanced at all their faces. Mom looked pale and worried. Adrian and Jeff wore matching fake smiles.

"What's going on, Mom?"

"Come in, honey," she said. "Let's talk inside."

"You're a good kid," said Jeff, patting my shoulder. "Everything's going to be fine. Listen to your mother and Adrian. All they want is what's best for you."

"Okay," I mumbled, staring up at him.

"Good man." Jeff nodded and walked away down the hall. He didn't look back.

"Come on Jason, let's go inside. Your mom's waiting."

Adrian's forced smile was gone, replaced by a more genuine look of smug satisfaction.

I hustled past him and into the living room. My mom motioned for me to sit down.

"We've found you a new school, Jason," she said by way of an explanation. "Or rather Mr. Montgomery has." I must have looked baffled because she inclined her head toward the door. "Jeff Montgomery. The specialist."

"It's called Presa River," said Adrian, sitting on the couch beside my mother. He took her hand in his and gave it a gentle squeeze. "It's a special school for, you know, *confused* kids. Jeff said a space has opened up there, luckily for us. They can take you on Saturday."

"Saturday? This Saturday? But school doesn't start for another four weeks." So many questions were flaring in my brain, but at first none would fully crystalize. I'd been in the sun all afternoon, and my head was throbbing and fuggy. "Where is this place? I've never heard of Presa River. Mom?"

"It's in downtown San Antonio." She couldn't look at me. "It's a residential school where they specialize in helping teenagers like you."

"Residential? You mean like a boarding school?"

My mother nodded and Adrian smirked.

"You bastard," I said to him. Then I turned back to my mom. "I don't want to go, Mom. I know I have to repeat eighth grade, and I'm sorry. But don't send me away. Please. I can do it here. I'll work harder, I promise."

"It's not just about your grades. This place can really help you," Adrian began.

I wasn't listening. I ran to my room, slammed the door, and threw myself onto my bed. I sobbed and shrieked and punched my mattress.

"Jason, please!" My mom said from outside the door. "This has nothing to do with Adrian. Your father and I decided it would be for the best. You need specialists. We don't know what to do with you anymore."

*

Three days later I entered *The Presa River Secure Center for Troubled Young Adults*. My mother handed me over to two 'teachers' who guided me into the center. The door we passed through was heavy, and it clicked shut behind us. I glanced back. There was no handle on the inside, only a keyhole; like a front door hung backward. Anyone could get in, but no one could get out without the key. I had hoped for a last glimpse of my mother through the wire-meshed glass in the top panel of the door, but she'd already scuttled away.

The hall I was led down had no windows. Florescent strip lights buzzed overhead, casting shaky white light over gray carpet, mint green walls, forest green doors. I wondered if the color scheme was supposed to be calming. If it was, it wasn't working. My heart pounded and my mind raced with a suffocating sense of déjà vu. I knew this place, or somewhere just like it. I could feel despair cowering behind each and every door.

My mind reeled, and in memory or premonition, I couldn't tell which, the hall darkened and decayed, filling my nostrils with the stench of mold and rot. I rose, floating in the putrid air, simmering in hatred. Then a defiant howl at the back of my mind returned me to the present with a jolt.

"You okay?" asked one of the teachers – a woman with greasy long hair and oval, green-rimmed glasses.

I wondered if she was joking. "Not really," I said. "I'm in a fucking prison!"

"Don't be so dramatic," said the other teacher. He spoke like a drag-queen, gruff, with an affected femininity, and he looked like an old-fashioned strong man. Bald head, bushy moustache, flab covered muscles and a potbelly. "This is a great place. Once you get used to it."

He knocked on one of the doors. All was silent on the other side, so he opened it and gestured for me to go in. Inside, were two beds, two wardrobes, two chests of drawers. Off to one side, was a half-open door leading to a white tiled bathroom. One of the beds was neatly made. The other was a swirl of clothes and blankets. A small boy with dense curly black hair perched within this nest like a frail bird, too weak and scared to fly.

His listless green eyes found mine. "Bulsara's wrong. You don't ever get used to it. Not if you're locked in. He gets to go home at night. And he gets paid to be here. That's why *he* thinks it's great."

"Come on Patrick, there's no need for that. This is Jason, your new roommate. Maybe you two can get to know each other. It might cheer you both up." Patrick and I exchanged a look, and Mr. Bulsara chuckled. "Or maybe not. But you could show him the ropes and make sure he doesn't miss lunch."

The teachers backed out and closed the door, leaving me alone with Patrick.

"That's your bed. That's your wardrobe. Those are your drawers. I didn't mean to freak you out. I just hate those do-gooding bastard teachers. Lunch is at one. Wake me up just before and I'll show you the way to the cafeteria. Good to meet you, Jason. I'm tired now, but we can talk later if you want."

Without another word, Patrick burrowed back under his blankets. I unpacked my stuff and waited, cursing my mother and plotting my revenge against Adrian.

At five to one, I called Patrick's name. There was no response so I said it again, louder. I got up and gave his shoulder a little shake, and he sat bolt upright, eyes open but still swimming back from sleep. His expression was one of raw terror, mixed with confusion as his surroundings came into focus. Once lucid, he sighed, and said, "Lunchtime?"

The cafeteria was a large room with windows all along one wall. Outside was a scruffy courtyard penned in by concrete and glass. Bushes and weed-choked flower beds, punctuated by low wooden benches, surrounded a brass sundial on a slender marble pedestal. It was a puny centerpiece for the size of the courtyard. Pointless too. The building was too high to allow the sun's rays to ever touch it.

Instead of joining the food line, I went and stood by the windows and gazed out at the greenery. There was a door in the corner. I tried the handle. It was locked.

"You won't escape through there," someone said from behind me. The voice was husky and musical, and female. "Not since they removed the fountain after the last one got away."

I turned. "Got away? I wasn't trying to-" My saliva turned to glue and I couldn't finish my sentence. She was beautiful. Her soft brown eyes sparkled, crinkling at the edges as she smiled.

The girl with the eyes and the voice ran a hand through her short, boyish black hair. "Why not? Most of us in here are escape artists. It's why we're locked up. And why they count the silverware after every meal."

"They what?"

"You're cute." She smiled again, then turned away. After a couple of steps, she stopped and looked back over her shoulder. "I'm Gwen, by the way," she added, before continuing onward to one of the corner tables.

A swarm of butterflies filled my stomach, and, in a trance, I shuffled away to get my lunch.

I *had* to watch her, but didn't want to gape like a moron. Instead, I just stole glances while shunting fries around my plate.

She caught me peeking at her and smiled. I blushed and looked away. Embarrassed, I didn't dare resume the game too soon. For five excruciating minutes, I forced myself to eat and study the clock face across the room. When I finally allowed myself another glimpse, she was gone. A boulder of panic fell into the pit of my stomach, scattering the butterflies, but not for long. I was bound to see her again, and soon; she couldn't leave any more than I could. All of a sudden, being locked up didn't seem so bad. I was dancing inside, freer than ever.

It soon became apparent the kind of troubled teenagers housed in our section of Presa River weren't the dangerous or criminal sort. Our group was made up of trauma victims and chronically depressed suicidal types. I was one of the milder cases, but like everybody else in there, I didn't see how being locked up was supposed to improve my life and bestow a more positive outlook.

I often wondered if any of the professionals who sentenced kids to the center realized how depressing it was for a depressed person to spend each and every day with other depressed people. I asked Mr. Bulsara about it once, and he told me it was better to keep us all together than to risk us being out in the world alone. Better for whom, he didn't say.

One thing the center did inspire in us all, however, was a healthy hatred of the system. This uniting passion helped some of us develop a camaraderie and mutual compassion we'd never known before. It was us against them, no matter what. For someone as shy as me, this breaking down of barriers was an unexpected blessing. It meant I was able to do things I'd previously thought impossible. Like talk to pretty girls without worrying about them laughing at me.

In the evenings I hung out in the TV room, vegetating with everyone else. Books were scarce. Most novels, especially the kind I liked, were banned. Fantasy books, or anything else that inspired the imagination were considered an entry drug to despair and Satanism by those who set the center's rules.

In the TV room, we were allowed to watch documentary channels, approved movies, sporting events and music videos (I'm not sure how the latter had slipped under the 'subversive influence' radar, but was very glad it had). We each had a seat allotted to us, and moving or talking wasn't allowed. Fortunately, our evening supervisors preferred to gossip and drink coffee rather than monitor our viewing habits, so over time these strict rules became ever more relaxed.

Whenever Gwen came into the TV room she wanted to watch music videos. We were at an age where music was becoming important to us, so nobody ever minded switching channels to MTV or VH1. We all loved rock music, especially *Guns 'n' Roses*. When *Paradise City* came on we would go nuts and dance around the room. Collapsing at the end of the song into the nearest sofa or chair. This teenage version of musical chairs was one of our small acts of rebellion. None of us wanted to be told where to sit.

One evening when I saw Gwen coming, I changed the channel to

MTV before she had a chance to say a word. I was rewarded with a smile, and then a wide grin when the familiar strum and pick of Slash's distinctive guitar intro filled the room. Everyone jumped up to sing along, play air guitar or air-drums. By the end of the song we were whirling and pogoing like dervish punk-rockers. I made sure I was bouncing next to Gwen, and when the last chord hit, I fell into the couch beside her.

"You did that on purpose, didn't you?" she said, in mock accusation. "You must have some magic power that grants people their deepest desires."

"I wish," I said, shrugging.

I could see her staring at me from the corner of my eye.

"Me too," she said. "Why don't you tell me you can, even if you can't really." She laughed then, and I laughed with her.

We spent the rest of the evening watching music videos together in silence. When *Paradise City* came on again we didn't get up with everyone else. Instead, she gave my hand a tentative squeeze that sent goose bumps rippling over my skin. We held hands and gazed at each other until it was bedtime, when I floated back to my room feeling every bit as magical as she wanted me to be.

The next day Gwen found me in the courtyard after class. Not everyone took advantage of the one hour of fresh air we were allowed before homework/counseling time began, but I never missed it. I would sit in the only corner the sun reached and squint into the bushes, pretending I was far away, out in some expansive wilderness, without a wall or locked door in sight. On this particular day, however, I merely basked in the sun's rays and daydreamed about Gwen.

"I thought I'd find you here," she said, appearing beside me as if summoned by my reverie.

"Hey, Gwen. You were looking for me?" I felt so foolish and shy.

"I was. Nature-boy."

I smiled. "My sister used to call me that. Among other things."

"I wonder why?" She grinned back. "So, is this your alone time? Or do you want to hang out?"

"It was. But I think I've had enough of being alone now."

"Me too," she said, reaching for my hand.

The sleeve of her shirt rode up, revealing raised crisscrossed scars on her wrist and forearm. She caught me looking at them and pulled her sleeve down, pinning it with her pinky finger.

"You think they're ugly," she said, catching me watching her.

"No. I just wondered what happened."

"Nothing happened. I did it myself." When I didn't say anything, she lowered her eyes and continued, "Actually, my uncle happened. I told my dad about it but he didn't do anything. Said we shouldn't cause trouble in the family." She shook her head and looked away, eyes glistening. She nodded at her forearm. "So I decided to escape."

"Escape? Oh, *escape*. I see." I squeezed her hand. "But you have now. Escaped, I mean. At least in here you're safe from him, right?"

She snorted. "Oh yeah. Apart from every night I feel him on top of me. In me. I smell his breath. The rum. The cigarettes. I wake up crying. In pain, real physical pain. Stinging, like I always did."

"I'm sorry," I said. It felt inadequate. It was inadequate. But I didn't know what else to say.

"Me too. I just wanted you to know. I feel like I need some sweetness in my life. I don't want to feel so crappy all the time, you know?"

I nodded. I did know. Everyone in the center knew. Some with more reason than others. Gwen was definitely one with more.

"What's your story Jason? Why are you in here?" she asked, squeezing my hand back and leaning into me.

I shrugged, feeling like a fraud. I had nothing to tell.

"It's okay, you don't have to tell me. Whatever happened, I'm glad you're here. You're sweet."

She kissed my cheek. My heart skipped and ached all at once. I wanted to smile and kiss her back, but couldn't. I'd never kissed anyone before and didn't want to mess it up. "You're sweet too," I said, putting my skinny little arm around her shoulder.

She nestled her head against my chest. A single tear rolled down her cheek. "I'm not," she whispered. "I'm ruined. And if I ever decide to escape again, I don't want you to be angry or blame yourself, okay?"

I patted her shoulder awkwardly. "You're not ruined," I said, unsure what else I could say. "You're perfect."

Gwen was my first love, and we latched onto each other in a big way. Unfortunately, the staff at the center discouraged any kind of deep connection between students. Roommates were changed weekly, and boys and girls were kept apart as much as possible.

The only times Gwen and I could be together was at meals, during outside hour, and in the evenings in the TV room. Our relationship went as far as holding hands and stealing kisses. Which was fine. I think it was the kind of innocent, loving relationship she'd been craving. Having the staff watching us all the time gave us a good excuse not to take things any further.

As our relationship blossomed, the only thing I was concerned about was losing my daily hour of peace in the garden. I didn't have to worry though. Sitting outside in silence was even better with Gwen beside me.

She told me the courtyard was the only place where she was able to contemplate a future beyond the center and her old life. She'd never gone there much before we got together, but now she loved to sit outside with me, and dream. I was glad the fresh air and flowers affected her the way they did me, and that we held that in common. I could never understand why the center only permitted students one hour outside a day. It was as if they wanted to ration hope.

My mother came to visit every two weeks. Again, center rules wouldn't allow her to come more frequently. I never looked forward to seeing her. She would tell me how well things were going outside, about her job, and how she was thinking of moving in full-time with Adrian.

In the beginning, I would tell her how much I hated it in the center and plead with her to take me home. Then I would sit, mute and surly, while she explained again and again how being there was in my best interest.

When Gwen became my girlfriend, I stopped begging to be taken out of the center. My resentment toward my mother didn't lessen, however. If anything, it intensified. Now, not only was I angry about being put there in the first place, I was also furious about having my time with Gwen interrupted so I could hear about how wonderful life was on the outside without me.

Gwen only ever had one visitor: her father, whom she always refused to see. She couldn't understand why I kept meeting with my mom, knowing how upset it always made me. I told Gwen I wanted my mom to see my rage and hatred as much as possible, to punish her for locking me away. And maybe I did. But, truth be told, it was also because I loved and missed her and didn't want her to forget about me. Eventually, my mother did intervene to have me released early. But not before I'd had my heart broken.

*

On a cool evening in late spring, Gwen and I were snuggled up on one of the stiff-backed couches in the TV room, whispering jokes and giggling while everybody else watched *Star Wars*. It was a night like all the others. A minute before bedtime we kissed, not breaking off until the moment the overhead lights were switched on.

She looked at me the same way she always did; a conspiratorial smile on her lips. Then she winked at me. We'd gotten away with it again. For all its innocence, love for us was something forbidden, a shared secret, fragile and precious.

When we said goodnight, I noticed Patrick staring over at us. For the first time since the week I'd arrived he was my roommate again. We still hadn't spoken much, and I wondered why he was so interested in Gwen and I. His expression gave nothing away. Those dull green eyes could have been looking right through us. Perhaps they were; he didn't acknowledge my raised hand when I waved to him.

Sometime after lights out, Patrick spoke, asking if I was still awake. Patrick loved sleep more than anyone I'd ever known. Including me. The small class sizes and short lessons at the center had cured my habit of falling asleep during class, but they hadn't stopped Patrick. He said he could sleep on a clothesline, and I didn't doubt it. So if he wanted to forego a fraction of his nightly rest to say something to me then I figured it must be important.

"Yeah. What's up?"

"Do you know why you're here, Jason?"

"What do you mean? Why I got put in here?"

"No. That's not what I'm asking. I don't want to know their reasons. I just wondered if you knew yours."

The question was confusing, and for some intangible reason it disturbed me. I laughed to cover up feeling spooked. "Do you know why you're here?"

"I think so. I mean, not completely, not all of it. But I'm working it out."

"Yeah? So what's your story?"

"You sure you want to hear? You might get nightmares."

I laughed again. "After living here nearly a year. There's not much left that could give me nightmares."

He was silent for a long time, and just when I thought he must have fallen asleep, he began his tale. His voice sounded odd. Like he was reading aloud, or in some kind of trance.

"I'm a failed escape artist like Gwen. Like most of the kids in here."

My skin prickled at the mention of her name, and his use of the words she used to describe suicide.

"Unlike Reece, whose bed you got when you first arrived, or that guy Teddy who hanged himself at Christmas, I've never been able to pull it off. Obviously, as I'm still here talking to you. Now, I don't think I'll ever do it. I've stopped trying. I don't believe suicide's my way now after all."

"I'm glad to hear it. But why did you think it should it be?" Even though Miss Sanchez had marked me down as suicidal on her assessment, I'd never seriously considered taking my own life. For one thing, I didn't think I was brave enough.

"Three years ago, I was playing hide-and-seek with my little sister, Kathleen," Patrick began. "She was eight years old. I was eleven. I hid under her bed and pulled her stuffed-toys in behind me. To cover me up. So she wouldn't see me. I left a space by my eyes so I could watch her looking. Her bedroom door was open, and I could see clear through to the living room. My mom and dad were in there, watching TV. Kathleen checked behind the curtains, under the coffee table; any space big enough for me to hide in.

"Somebody knocked on our door and my dad went to answer it. I heard raised voices. Quarrelling. Kathleen stood still as a statue in the hall. Pulling at the hem of her Minnie Mouse t-shirt. My dad backed into view, and I saw the shoes of two men shuffling in after him. They were both wearing dirty white Nikes. One had red swooshes, the other blue.

"The men were jabbering at my dad, low voiced, urgent. I couldn't hear what they were saying. Dad told Kathleen to go to her room. She got as far as the door when the shot was fired and Dad collapsed. His chest was caved in and bleeding, his eyes blank and staring. They'd killed him. Mom screamed and threw herself at one of the men. Another shot, another spray of blood, and she was dead, too.

"Through it all Kathleen didn't make a sound. She just stood in the doorway, watching. Pulling at the hem of her shirt. One of the men kneeled next to my dad and started going through his pockets. The other bent down and said something to Kathleen. She shook her head, then he stood, raised his gun, and shot her as well.

"I stayed under the bed for a long time. Staring at the mush of pink gunk, white bone and curly hair that used to be my sister's head. When the police found me and pulled me out, I had this crazy idea she might still be alive. I kept asking them to save her. There wasn't a drop of blood on her t-shirt.

"A few foster homes and a couple attempts to reunite with my family later, and I ended up here in Presa River. Times gone by and I've managed to figure a few things out. The main one being now's not my time to die, no matter how much I might want it to be.

"My sister saved me, you know that? The last thing she did before they killed her was save my life. That's why she shook her head. The guy was asking if anyone else was in the house. I'm sure of it. Now, how can I throw away the last gift she gave me? What kind of brother would I be if I did that?"

"Shit Patrick, I don't know what to say."

"I know. It's okay, Jason. It's been three years. Life goes on for the living. If for no other reason than as a tribute to the dead who got them there. Speaking of which, I need to sleep. Sometimes when I dream, Kathleen finds me under the bed and everything's okay again."

I was weeping, but tried to keep my voice steady. "Okay, Patrick. I hope she finds you tonight. Night. And thanks for telling me."

"Night Jason."

I lay there listening to Patrick's breathing slow into the deep rhythm of sleep. It amazed me how easily he could drift away after all he'd been through. Then I realized. Dreams were his home, his solace. The same way forests and lakes were mine. I squeezed my eyes shut and imagined walking among tall moss-covered trees. I saw moonlight on still water and Gwen beside me.

*

The screaming penetrated into my sleep, jolting me through the surface to consciousness, and all at once, I was awake and alert. I switched on my bedside lamp. The shrieks were interspersed with moans, wails, loud gulping breaths. The commotion was coming from outside. Somewhere down the hall.

"Patrick? You hear that?"

Silence. Patrick could sleep through the New Year's Eve countdown in Time Square. I put on my slippers and went to investigate. As soon as I stepped into the hall, I almost ran straight back to bed. A nightmare I could handle, reality was a different thing. Doors were opening left and right. Sleepy kids peered out. Staff members were running in and out of a room down the hall.

I ran to the room, too. Even knowing what I would find there, I still ran. I had to see her.

When I reached Gwen's room the screams coming from inside had subsided into heaving sobs. The sound was primal and heart wrenching. I pushed past someone. I don't know who it was. A creased poster of Axl Rose sneered down at me over the messy tangle of covers on the empty bed. Had she had a nightmare? Clarisse, Gwen's roommate, was hugging her knees and rocking back and forth on the other bed. A teacher sat with her, cooing soothing words. Neither noticed me. The carpet was dark and wet, squelching underfoot.

I found Gwen in the bathroom. Lying on the floor. Alone. A sheet had been placed over her. Halfway down her body, red flowers bloomed; vivid dark red roses on a canvass of pristine white. Pink wisps washed into the glassy puddle surrounding her. Blood smeared the walls. A crimson pool filled the bathtub.

She'd done it. She'd escaped. I collapsed against the wall and slid to the floor. Water soaked into my pajama pants and I touched the swirl of pink that ran from her to me. I wanted to say goodbye, but she'd already gone. My heart clenched. My breath came in tight gasps. The butterflies I felt whenever I saw her swarmed together and fused in my belly, becoming a leaden lump. I turned and pressed my forehead against the cold tiles. I pulled back my head and smashed it against the wall, over and over, feeling nothing but the pain of her loss. All else was numb.

*

In the weeks after Gwen's 'escape' I shut down. I didn't want to speak with anybody. The teachers tried to reach me. So did Patrick and a few others. But I was gone. I was with Gwen. I didn't want to eat, I didn't want to do anything; I just wanted her back. All I wanted was to follow her, and I hated myself for being too weak and afraid to try. In death, as in life, I let her down.

Gwen had cut her wrists with a blunt nail file. She'd run a bath sometime in the night, got in, and then dug and pulled at her veins until they'd ripped and spurted open. Then she'd laid back and let her lifeblood seep away into the water. I wasn't strong enough or determined enough to do anything like that. My motivation was misery. Hers had been something deeper and darker.

At first, I refused to meet with my mother when she visited. She persisted though, and eventually I agreed to see her. She'd heard what had happened from a member of staff and tried talking to me about Gwen. I was going through hell, and the thought of allowing the woman who'd had me locked up help me deal with the grief of losing Gwen was insulting. She talked at me while I stared at the floor, marking time, waiting for visiting time to be over.

"You do know our intentions were good, don't you?" she said, as she was getting up to leave. "We thought you being here would make everything better."

I shook my head in disbelief, and looked up into her face. "This was never for me, Mom. You know that. And now things are worse than ever."

My mother nodded. "Okay, Jason. I'm going to get you out. Today. I've decided. It *is* making you worse. I can see that now. How would you like to come back home? We have a big house out in East Creek now." She caught my look, and added, "He wants you home too. We all do."

The bell sounded and I stood up. My mom went to hug me, but I ignored her and left the room. Ever since arriving at the center, all I'd wanted to do was leave. At first I'd wanted to go back home. Then I'd wanted to run away with Gwen. Neither was possible anymore. Home was gone, and so was Gwen. Even the boy I'd been when I entered Presa River was gone.

I went to my room and packed my bag. Even if my mother was lying about getting me out, it was my turn to change rooms the following day, so I would be ready for that if nothing else. When I finished packing I sat on my bed and tried not to think about Gwen. The dull ache in my heart couldn't be ignored and I was soon crying again. Patrick knocked on the door before coming in to our room. He stood in front of me until I looked up at him.

"So it's true. You're leaving. Bulsara said you were."

I nodded.

"Good luck."

I wiped my eyes. "You too. I hope you figure out what you're supposed to do with Kathleen's gift."

Patrick smiled. He always smiled when I said his sister's name. "I will."

Bulsara poked his shiny head around the door. "Ah, there you are, Jason." He glanced from Patrick to me to my stuffed backpack. His mustache twitched. "You all ready to go then?"

I nodded and glanced at Patrick. "I think we all are."

Patrick chuckled. It was the first time I'd ever heard him really laugh.

"Funny guy. Okay, let's go then," said Bulsara, retreating into the hall.

Patrick pushed the door closed behind him and turned to me. "Hey Jason, I've been meaning to tell you this for a while but never found the right moment."

"Probably because you slept through them all. What is it?"

Patrick smiled again. "That's true, I probably did. Or maybe you were too busy feeling sad about Gwen to be approachable. Anyway, I just thought you should know something. You're not alone, you know. I'm not sure if that's a good thing or a bad thing, but it is true."

I smiled back at him. It was typical of Patrick to attempt to say something reassuring and have it come out creepy as hell.

Fifteen minutes later, I walked out of Presa River. The sky outside was big and blue and achingly beautiful. I'd been locked up for nearly a year for the crime of being an introvert with bad grades.

I hadn't been corrected. But I had been changed.

Indigo Summers

We passed a huge clock-tower on the way in to East Creek.

"Oh look," said my mother. "Time for a new start."

I wasn't sure if the joke was hers or one she'd heard and recycled for my benefit. Either way, I smiled; a new start was what we all needed.

East Creek was the kind of suburbia middle-class America, in their cookie-cutter homes, dream of living in. Imposing, individual town houses with football-field-sized lawns, and sidewalks wide enough for swarms of kids to ride their bikes up and down. It reeked of Saturday afternoon barbecues, car wax and pseudo-contentment. But it was a lot better than Presa River.

When we pulled onto the driveway, Adrian was standing waiting for us on the porch. King of the castle. The short sleeves of his button-down shirt flapped and fluttered around his matchstick arms.

Mom waved at him.

"Now Jason, I want you to be nice to Adrian." She spoke out the corner of her mouth. The other corner remained in a rictus grin of congeniality. "He's opened up his house to you, which wasn't easy after everything that's happened. But he's a good man, a forgiving man. Maybe you could take a leaf out of his book and do some forgiving, too."

"Sure. If it'll make you two feel better about having me shipped off to the loony bin for no good reason."

"Jason, it wasn't-"

"I know, Mom," I interrupted. "It wasn't Adrian's idea. You told me already. At least a hundred times. Seriously, Mom, the last thing I want to do is make waves. I mean, look at this place. If I can't be happy here, there must be something seriously wrong, right?"

"Right," she agreed, without the least trace of irony.

Adrian had done the traditional Saturday afternoon thing and burnt some pork ribs and undercooked some chicken legs. We sat on the back porch and ate the charcoal and salmonella with corn on the cob and mashed potatoes. Kendra put in a brief appearance before disappearing off to a football game with her girlfriends. I was glad to see her, and it was reassuring to know I wasn't the only one uncomfortable playing happy families.

Adrian's house was built on a slight rise at the edge of the development. It had commanding views of the surrounding country. There were meadows, a huge swathe of woodland, and a two-lane blacktop twisting between. A huge tree way out in the middle of the woods caught my eye. Judging by its shape and size I figured it to be an ancient pecan. I wondered how long it would be before the bulldozers moved in and flattened the woods where it had stood for so long. The next phase of super-size desirable homes had to be built somewhere, and that was prime real-estate right there.

"I have to go," I said, cutting through the small talk. "Thanks for lunch, Adrian." I didn't want my mother to say I wasn't at least trying to be nice.

"Where are you going, honey? Lunch isn't over yet."

I pointed to the woods. "Over there. I need to take a walk among the trees."

"But your mother made you a cake!" Adrian whined.

I glared at him. "I haven't seen a tree for nearly a year, Adrian. Cake we had every Friday night in the center."

Adrian's face screwed up and his ears reddened. He bunched the napkin in his hand and drew back his fist. I thought for a moment he was going to hit me, but instead he hurled the sauce-stained piece of cloth at my face. He was three feet away, but still missed. The napkin flew past my ear, unfurled, and wafted to the floor a foot or so behind me.

"You're so ungrateful," he said, stamping his foot for emphasis.

I shook my head, mouthed *sorry* to my mom and left. Being nice wasn't going to be as simple as I thought.

*

I reached the blacktop in less than five minutes. The woodland was nestled in a long wide valley and I could still see the big tree poking out through the canopy. I fixed my eyes on it and stepped out onto the road.

"Look out!" someone called from across the street.

I froze and car flashed by, missing me by inches. Heart pounding, I watched the car disappear around the bend, glanced left, then right, then sprinted over the road.

"Shiiiit, that was close! You okay?" Two boys were gaping at me from the edge of the wood. They were my age, or close to it. Geeks like me. One of them held a stick with a forked end.

"Hey, yeah, thanks to you. Do they always drive so fast around here?"

"Of course," said the boy with the stick. "It's called the death loop ain't it?" He had long, wispy brown hair and glasses. His friend was chubby and red-cheeked, with a spikey blond flattop.

"It is, huh?" I said.

"Don't you know anything?" said Flattop. His voice was breaking hard, oscillating in tone between Mickey Mouse and Barry White. I figured it must have been the other guy who'd warned me about the car.

"Not much," I admitted. "But I do know you're looking for snakes."

Flattop glanced at Long Hair.

Long Hair shrugged and said, "How do you know that? You psychic or something?"

I shook my head. "Why else would you have that stick? You don't look like the water-dousing types."

They smiled, recognizing a kindred dweeb. Long Hair was called Dudley. Flattop's name was Brandon. They said they were on their way back from hunting copperheads and cottonmouths by the creek. I noticed a length of sacking hanging limply from Brandon's back pocket, and nodded to it.

"Guess you didn't catch any then?"

"Nope. Saw quite a few, but we didn't catch them," said Dudley. I must have looked surprised, because he added, "They're real fast, you know. And dangerous. Not easy to catch at all."

"They are fast," I agreed. "You sure you saw both at the creek? Not often you find those two in the same place. Copperheads like to hang out around rocks and under fallen trees usually. But I guess since it's still springtime they might have gone down to visit the water moccasins at the creek edge. They are known to do that sometimes around this time of year."

"Water Moccasin?" squeaked Brandon, before clearing his throat.

"Cottonmouths," said Dudley. "It's another name for them. Everyone knows that." He squinted at me. "Sounds like you know a bit about snakes."

Snakes and lizards had fascinated me for as long as I could

remember. My friend Ellen had called me the snake whisperer because of my unerring ability to find and catch them. On top of that, the only book I'd read (and consequently re-read several times over) while I was in the center was called *The Worldwide Compendium of Reptiles and Serpents*.

"A bit," I admitted. "If you want I can show you how to catch them without being bitten. Just so long as you don't harm them afterward."

Brandon raised his eyebrows. Probably deciding communication with facial expressions was safer for him than speech right now.

"Course we don't want to harm them," said Dudley, looking me up and down with renewed interest. "You really think you know how to catch them?"

"Uh-huh. And I can teach you two how to do it, too. But there's something I want you to do for me in return. There's a big tree. I think it's a pecan. You can see it from all around-"

"The grandmother," said Dudley. "What about it?"

"Can you show me where it is? I know it's over there somewhere, but I don't know these woods well enough to go straight to it. I'd appreciate you taking me there after we're done catching snakes."

"Deal," said Dudley, and the three of us shook hands.

*

"*Agkistrodon contortrix.*" I whispered the Latin words like a magic spell.

Brandon and Dudley stared at me.

"Say what?" rumbled Brandon in a rich baritone. He grinned with relief and pride. One day his voice would sound like that all the time.

"Copperhead."

"Where?"

I pointed to a cleft in a tree, close to the ground near the edge of the creek. The copperhead was lying still. I plucked Dudley's snake-stick out of his hand and faded back, skirting around the trees to approach the orange and chestnut serpent from its blind side. Once in place, I made a quick jab with the stick and pinned the snake to the ground just behind its head. Its body whipped back and forth. Carefully, I replaced the stick with my free hand and lifted the snake into the air. I turned its gaping jaws toward my new acquaintances.

"Whoa!" they chorused. Brandon rushed forward and thrust the sack at me.

"You want to keep it?" I asked him.

He shook his head. Smart boy. Copperheads are highly venomous and don't make good pets.

"We just wanted to get a good look at one," said Dudley, arriving at Brandon's side and leaning in. He studied the now subdued serpent, then glanced up at me and grinned. "Wow, that was so cool!"

We stood there admiring the snake for a few minutes and then I let him go. The next few hours were spent finding and catching a variety of non-venomous snakes to give the guys some practice. At some point, another of their friends, a tall red-headed guy called James, caught up with us and joined the hunt.

It felt so good to be out in the woods again, and as nice as it was to find some new friends with a common interest, I was itching to get away by myself. All I could think about was climbing the huge tree I'd seen from Adrian's house.

"Hey, can we go see the grandmother tree now?" I asked Brandon after he'd successfully caught and released his third garter snake.

James pulled a face. "Not sure about that, sport. Aren't you Kendra Pednault's brother?"

"Yeah, so?"

"So, I heard you were a bit crazy. I don't want to be responsible for you jumping out of a tree."

"Why would he jump out of a tree?" asked Dudley.

"Two words," said James. "Presa. River."

"Isn't that where they put the hard-core criminal kids?" said Brandon, his voice soft and even for a change.

"And the crazy, suicidal ones," said James. "He's a bit too nerdy and weedy to be a bona-fide delinquent, so, he's either cracked in the head or has a death wish. Probably both. Care to enlighten us, Jason? What kind of crazy inmate were you?"

James folded his arms over his chest, waiting for my reply. Tears formed in the corners of my eyes. I blinked them back. In the darkness behind my eyelids I saw a damp white sheet, a bloody handprint, Gwen's face; cold, waxy, dead. I wanted to cry and scream and smash my fist into James's smug face. But I did nothing. I just stood and stared back at him until he looked away. When he did, I noticed the other two were gawking at me as well.

Dudley sighed and pushed James aside. "He's the kind who wants to breathe fresh air and climb a big tree now he's out. Come on, Jason. I'll take you there."

I followed Dudley along the trail, while James and Brandon stayed behind. I could feel their eyes burning into the back of my head as I walked away. It seemed making friends was going to be just as hard as it had always been. So much for a new start.

When we reached the grandmother tree, I forgot my embarrassment, the center, and everything else. I even forgot Gwen and the aching hole she'd left in my heart for the briefest moment. The giant pecan was at least a hundred feet high, with thick, evenly spaced branches.

"Can you give me a boost up to the first branch?" I asked Dudley.

He gave me a searching look. "You're not-"

"No, I'm not." I smiled. "As rough as things got in the Presa River, I never wanted to kill myself. And now, after everything that happened in there, I'm the last person who'd ever consider it."

Dudley smiled back. "Well, whatever happened, you're out now. I'm glad you managed to escape."

I flinched at the last word, but he didn't seem to notice. He braced his back against the tree and clasped his hands together for me to step on to.

I stepped up, grabbed the lowest branch and pulled myself up into the tree. As I climbed, a spark lit inside me and all thought and feeling fell away. It was just me and the tree. I used my feet and hands, chest, neck, elbows, knees. I twisted and stretched, reached and jumped. When I climbed higher than the surrounding trees, I felt the light shift and peered through the leaves. The sky was open; bright and blue and clear, stretching forever above the sea of green treetops. I leaned back against the wide trunk, my legs dangling either side of a sturdy branch. Now I was home.

After a while, Dudley called up and asked when I was coming back down. The others had arrived and were getting bored. James hollered an apology. I thanked him and told them all to go on home. I wanted to stay where I was.

After kicking around the bottom of the tree for a few more minutes, they shouted their goodbyes and left. I stayed up in the grandmother tree until sunset, and made my way down in the dark.

*

To my surprise, the time I'd spent at Presa River made things easier for me when I started back at school. The teachers had been informed of the tough time I'd been through with Gwen so they cut me a lot of slack. I could get away with my old habits of being surly and not turning in homework without fear of consequence, at least for a while.

The kids all knew where I'd been, too. So, with them, I had the mystique and prestige of a badass. A geeky, tree-climbing, snake-hunting, fantasy book reading badass - but a badass nonetheless.

Dudley and Brandon were high school freshmen like me. James was a sophomore. Outside of school, the four of us would hang out in the woods together, ostensibly looking for snakes, but mostly just goofing off. I enjoyed having friends, but I was still a loner by nature. Most days, I would lose the others at some point and go climb the grandmother or one of the other big trees by myself.

It wasn't long before I discovered the reason we hardly saw any other kids out in the woods: it had a sinister reputation as a place where children went missing. Rumors of satanic cults and murderous sacrifices in the area went back years. As a consequence, most people were too afraid to go anywhere near the trees.

On our daily adventures, we would occasionally find animal heads mounted on stakes, blood smeared on rocks and over the doorways of hidden shacks. Each new discovery gave us a thrill of fear and excitement. Although deep down we assumed it was just somebody messing around, or keeping the legends going in order to maintain the quiet and peace of the woods, there's no denying these disturbing finds added to the primal appeal of the place. We couldn't get entertainment like that on the sterile streets of the East Creek development.

One time, when exploring an old cabin, we found a cardboard box full of maggot-ridden horseflesh in the middle of a pentagram scratched-out on the floor. We thought we'd found some poor murdered kid's remains and decided to embark on a heroic quest to try bring justice to his killers. We took the stinking box of meat, buzzing with flies, to the local police station, where it was dumped into a garbage can after a cursory analysis by one of the mustachioed sheriffs.

"I know horsemeat when I see it," he informed us. "Horse-shit, too. I've seen, heard and smelled enough of both to last three lifetimes. Now you best get on your way before I charge y'all with wasting police time."

In light of the sheriff's terse warning, we felt it wise not to alert the authorities when we accidentally set fire to a meadow the following week. The resultant blaze threatened not only our beloved woods, but also the entire East Creek development, according to local news.

In everything we did, except the snake hunting, James took the role of ringleader and decision maker. He was good fun, but the two of us were never close. He resented my knowledge and skill with snakes and the fact I preferred to stay up in trees rather than follow his orders. For my part, I resented him because he gave orders. But as I had no aspirations to lead our little gang, I let him dictate most of our mischief, safe in the knowledge I could withdraw and be by myself if I ever felt bored or frustrated with him.

The one time we did form a close bond was the day he nearly died. It was also the day our friendship ended.

*

The four of us were cutting along a trail through the high grass beyond the woods on our way to a local swimming spot. James was leading, I was next, with Dudley and Brandon lagging behind. A snake slithered out of the grass and onto the trail about twenty feet ahead. It stopped and I peered around James to get a better look at it. Recognition dawned, and I grabbed James's arm and pulled him back.

"We can't go any further. There's a big rattler up ahead." In the corner of my eye I saw the snake dart away, disappearing into the long grass again.

James yanked his arm free and turned back to the trail. "Where?"

"Up there. You didn't see it?"

"You're full of shit, Jason."

"What's going on?" asked Dudley. He and Brandon had just caught up with us. "We going swimming or not?"

"I am," said James. "I think Jason wants to climb trees. He *says* there's a rattler up ahead."

"Not just a rattler, a big diamondback."

"You sure?"

I nodded.

"We better go back then," said Dudley. "We can always walk around the field. It's not worth the risk going on."

"What risk?" James scowled at me. "This loser says there's a snake and you all pee your panties. Take the long route if you want. I'm going straight to the water-hole. You guys can bake your brains wandering around in the sun. When you get there, I'll let you know if I found Jason's diamondback."

"I'm serious, man. Don't go up there," I said, making a last effort to stop him.

James wiped his brow with the bandana he kept in his back pocket. "Phew, sure is warm out here. Sun and steaming bull-crap, what a combination! See you at the water hole, suckers." He turned and strolled away, whistling.

Dudley shrugged and he and Brandon made a move to head back.

"Wait. Let's just see if he makes-" Before I could finish my sentence the whistling was usurped by a high-pitched scream.

"Sounds like he found it," said Dudley.

James was curled up in the dust clutching his leg. I glanced around, cursing myself for not bringing my snake-catching stick.

"Get it, Jason! It bit me! Don't let it come back. Oh Jesus, shit. It got me, man. Please, you got to help me!" He was crying and shaking, already going into shock.

I spied the snake a few feet away, coiled and rattling its tail in warning. It wasn't like any of the diamondbacks I'd seen in the area before. It was bigger and the markings were different. But it *was* a diamondback. I knew it from the reptile book at the center.

"It's okay," I said to it. "I'm just going to take my friend away and leave you in peace." I kept my eyes on the snake as I took hold of James's arms. Brandon appeared at my side and helped me drag James back. "That's right. We're going now," I continued. "You're a long way from home, huh? No wonder you got scared. It's okay. We're going. We're not going to bother you anymore."

"Stop talking to the damn snake will you! This is your fault. You knew it would bite me."

"He tried to warn you, you idiot," said Brandon.

We managed to haul James a safe distance from the serpent, but the question now was what to do next. As if reading my thoughts, Dudley took control of the situation. "I'm the fastest runner. I'll go for help," he said. "You're the snake expert, Jason. You've got to keep him alive until I get back."

I nodded, and Dudley sprinted off toward the woods.

"Suck it out," said James. "I'm going to die if you don't get the poison out."

I used my penknife to cut back the bottom of his jeans. The bite was a few inches above his ankle. The puncture wounds were reddish-black and puffy, and the surrounding area was already swollen and horribly discolored. "You're going to be fine," I reassured him. "It's not too bad. Sucking the poison out only happens in the movies. In real life we have antivenin. Dudley's gone to get some. Right now, all you have to do is lay back and stay calm."

"Stay calm? I've just been bitten by a fucking rattler!"

I squeezed his hand. "That's why you have to stay calm. If you get too agitated the venom will get further into your system. You need to be cool. It'll slow it down."

Brandon peered over my shoulder at James's leg, and then removed his t-shirt.

"Good idea." I snatched Brandon's shirt, then removed my own and laid them over James's chest and legs.

"I was going to make a tourniquet with that," said Brandon, looking indignant. "You know, so the venom doesn't spread."

I shook my head. "This is better. He needs to be covered more than he needs to lose a leg through lack of circulation." Brandon's brows furrowed but he didn't say any more. I looked at my watch. Dudley had been gone three minutes. It already felt like an eternity. "Dudley's going to be back soon," I lied. "Just keep calm and you're going to be okay."

I laid down in the dirt next to James and kept him still. I held his hand and whispered words of comfort and encouragement. He gripped my hand back until he was too weak to be able to anymore. Throughout it all, our eyes were fixed on each other's. He begged me not to let him die and I promised him I wouldn't.

It took half an hour for Dudley to arrive with help. We heard the ambulance pull up in the distance, and within a few minutes a team of paramedics and police officers had swarmed over the long grass toward us. By this time James was in a bad way. He was bleeding from his mouth and his pulse was faint. The snakebite was angry and bloated, oozing thick dark blood. The paramedics put him on a stretcher, injected antivenin, attached an I.V. and carried him away.

"When you boys gonna learn not to mess with things you don't know nothing about?" It was the sheriff who'd disposed of the horseflesh and given us the lecture the previous fall. He spat a glob of tobacco-stained saliva on the dirt.

"Go easy, Bob," said his partner. They both had the same drooping blond mustaches and squinty eyes. I wondered if they might be brothers. "Paramedic said these boys probably saved the kid's life."

Brandon cleared his throat, steeling himself against recalcitrant squeaks. "He's gonna make it then?" he asked.

"Maybe," said Bob's partner. "Maybe not. If he does it'll be thanks to you boys acting so fast and doing such a fine job of looking after him. If he doesn't, it'll be due to the rattler. Ain't no more to it than that. Speaking of which, where is the damn thing?" He took his revolver out and aimed it into the long grass. "We need to verify what kind of snake it was bit him."

"It was an Eastern diamondback," I said. "You don't have to look for it. It'll be long gone by now anyway. Besides, I already told the paramedics. The markings are pretty distinctive, and it was too big to be a Western."

"We don't get any Eastern diamondbacks here in Texas," said Bob.

"They have them down in Florida, I think," said his partner.

"All through Georgia, Alabama, and Mississippi, too," I agreed. "And now we have at least one here in Texas. I'm sure of what I saw, sir. I know snakes."

"And you're the one who took care of the boy when he got bit?"

"Yes sir."

"Hmmm, okay." Bob looked out over the field. "Give Larry your names and addresses. When you're done with that you can tell me what happened from the beginning." He turned to Larry. "Maybe we should call Thelma. Get her out here to find the darn thing. Then we'll see if this little smartass knows as much about snakes as he thinks he does."

*

Officers Bob and Larry came by the next day. They told me the renegade snake had been caught, and that James was going to be okay. They also said my identification of the snake had been correct and reiterated that without my help James would almost certainly have lost his life.

Unfortunately, that's not the way James saw it. In spite of everything, he still blamed me for what had happened. He cut me off and did his best to make sure Dudley and Brandon did the same. After that, I spent more and more time alone in the woods. I didn't miss the company. And even if I had, it wouldn't have mattered; I was about to be headhunted for my first proper job.

*

I was on my way down through the familiar branches of the grandmother tree, when I heard a sound like a maraca beating out a slow rhythm close by. I followed the noise to its source where I discovered a beautiful iridescent black serpent with a shining pink underbelly. It was draped across a large rock, casually devouring a rattlesnake. All that could be seen of the rattler was the end of its tail, jerking in time to the contractions of the larger snake's belly.

I sat back on my haunches and watched, mesmerized by the sound and the spectacle of the Texas indigo slowly sucking up its deadly prey like a string of fat spaghetti. This was *the* snake I'd been hoping to find all summer. It would be the prize of my collection.

Much to my mother's and Adrian's dismay, I often had as many as twenty-five serpents of various sizes crammed into the glass tanks and steel cages littered around my bedroom. Most were Sonoran king snakes (which I bred to sell at a local pet store), but I also had a colorful collection of racers, garters, hognoses and a variety of water serpents. So far, though, I hadn't seen one Texas indigo in the area, much less been in a position to catch one.

I soaked in the moment. Enjoying the show while waiting for it to finish its meal. Once it was done, it lay basking in a thick ray of sunlight. Shimmering rainbow hues glistened across oil-black scales. I was torn between seizing it at once or leaving it where it was so I could admire it a little longer.

A sudden movement a few yards beyond the snake caught my eye. There was somebody there. A thin man with close-cropped curly hair, a pressed khaki shirt and garish green-framed glasses. He saw me looking at him and indicated for me to stay where I was, miming that he wanted to come over and speak with me.

He took a circuitous route to avoid the snake and arrived at my side a few minutes later. I gave him a nod of acknowledgement, then did a double take; he was actually a she.

"Majestic, isn't he?" she said. Her voice was gruff - a smoker's voice

- but definitely female. "You want to catch him, don't you? I can see it in your eyes. Your face was all lit up watching him just now."

I frowned and kept my eyes on the snake. I didn't want it to disappear while this weirdo distracted me. I'd waited too long already.

"I've been looking for an indigo, too," she said.

I shot her a questioning look and she smiled. She had baby teeth and thin lips. Her muddy, yellow-flecked eyes were set a little too far apart and were magnified by fishbowl lenses framed in moss-green plastic ovals.

"Don't worry. He's still yours. Even if I did see him first. Him and the rattler he was stalking."

That got my attention. "You saw it catch the rattler?"

"Sure did. Made my day. That and finally catching up with the other indigo I've been hunting."

"You found another one?" I asked, incredulous.

She laughed. "There are actually three indigos here right now. The black beauty enjoying the sun over there, me, of course, and you. You are Jason, right?"

"Who wants to know?" I asked, glancing sidelong at her.

She chuckled. "Calm down. Your prize will sense your distress and slip away if you're not careful." She patted my back. "Let's start again. My name's Thelma Fraser and I'm a herpetologist; a snake hunter like that indigo over there. I'm looking for someone to help me with my work. Someone with knowledge and a passion for serpents. Another indigo, if you will. You want the job? If you do, you best stop gawking at me and show me you got the chops." She nodded in the direction of the snake. "Be quick though. He's got a head start on you."

I looked back just in time to see the tip of its shining tail disappear behind the boulder. Snatching up my snake stick, I rushed across the clearing and pounced. Fortunately, it was still deep in the torpor of after-dinner lethargy and I caught it with ease, depositing it in the burlap sack I carried with me. I presented the writhing mass to Thelma, feeling pretty good about my display.

She smiled and handed me a business card. "Not bad. Now you better take him home and settle him in. My address is on the card. Come by tomorrow morning at seven-thirty. You have a lot to learn before you can be useful, but I'll start paying you from day one."

I looked at the card, then back up at her. "Okay, I will. Thanks."

"See you tomorrow then." Without another word, she got up and stalked away through the bushes toward the main trail.

"Hey Thelma," I called after her, holding up the bag with the snake in it. "How do you know this snake's a he?"

"You mean to tell me you can't tell the difference?" She called back, without turning around. I flushed bright red anyway. "You wouldn't be the only one with that little problem, young man. Seven-thirty tomorrow. Don't be late."

*

Thelma lived in a ranch-style house, set behind a large hedge and surrounded by a threadbare lawn. The path to the front door was cracked concrete. I rang the doorbell and tapped the snakehead doorknocker at twenty-five after seven. She didn't answer. I could hear moving around inside so I knocked again. After knocking twice more, I gave up and sat on the stoop to wait. Five minutes later, at exactly seven-thirty, the door swung open and Thelma told me to come in.

Nothing could have prepared me for what I found inside this unassuming neighborhood house. The moment I stepped across the threshold, an enormous lime-green iguana turned its head and eyeballed me from the banister of the stairs. Scores of aquarium tanks lined the walls, stacked one on top of the other, floor to ceiling. Darting, slithering movements seem to emanate from every dark corner. The buzz and flicker of florescent lighting reminded me of the center. I pushed the thought aside and peered into one of the brightly illuminated tanks to see what it held.

I counted five tarantulas, crawling around in their little artificial habitat. An African bird-eating spider occupied the tank above. Below, a black-widow sat, alone and shining in mourning black. I backed away so I could take in the rest of the tanks and almost stepped on the iguana who'd descended from its perch to take a closer look at me. He skittered away and I bashed into a tank containing an evil-looking scorpion. The scorpion's tail struck the glass next to my face and viscous venom ran down the inside of the tank in a phlegmy smear.

"Leave that poor thing alone, will you? You're wasting all the venom!"

"I'm sorry. The big iguana startled me-"

"Which iguana?"

"A six-footer. How many do you have?"

"Ah, Max. He's six-feet seven-inches, actually. I would say that means he's bypassed big and is now as long as a basketball-playing-freak is tall. Don't mind him, he's been with me years. Ever since my first trip to the coast of Ecuador. Now, to answer your question, I have seventeen Iguanas in total, along with one-hundred and seventy-six other species of ordinary reptile. I also have around two-hundred and fifty spider species and nearly a hundred types of scorpion. My real passion, however, is serpents. They're through here. Do you want to see?"

For an oddball boy like me, this was heaven. Grinning, I followed her down the corridor under the watchful eye of the deadliest arachnids, insects and scorpions in the world.

We entered what would have been the living room in a normal house. Here it was set aside for the most astonishing collection of serpents I'd ever seen. It was as if the pages of *The Worldwide Compendium of Reptiles and Serpents* had been brought to life in front of me. The vivarium at the local zoo didn't even come close. I told Thelma that and she laughed.

"Yeah, I give them my cast offs. My ambition is to collect every species of serpent in the world. So, as you can imagine, space is at a premium. All the zoos in the United States have received snakes from me over the years. Speaking of which, we need to move your friend on. Could you bring that crate over here?"

I looked around and saw a perforated wooden crate with a sliding hatch roof. It was heavy, and I had to half-drag it to where Thelma was standing. She unlatched one of the high cages, stuck her hand inside, and withdrew the Eastern diamondback that had bitten James.

"You should thank this little beauty, Jason. She's the reason you're here. If you hadn't recognized her I might never have heard of you. Do you want to hold her while I make the crate a bit more comfortable for her journey?"

"Sure." I took the snake from her hands, keeping a firm grip behind its head. "Weren't you worried she might bite you when you put your hand in the tank?"

She gave me an odd look, made even odder by her wide-set and magnified amber eyes. "No, why would I be? I've lost count of the times I've been bitten by her sort. You know how feisty vipers can be. But I'm almost immune to their little kisses now. Besides, she's very sweet and we've become great friends.

"However, this is no time for sentimentality. Poor thing's going to Baltimore zoo where she'll freeze her ass off, which should teach her a lesson about wandering too far from her natural habitat. I already have a fine specimen of an Eastern over there next to the king cobra. I'm not fricking Noah and it never rains worth shit around here, so I don't need two of each."

We packed the diamondback in the crate and carried it out onto the porch, ready for collection. Back inside, Thelma gave me a lesson on how to handle the different kinds of venomous snakes, before taking each one in turn and showing me how to milk them for their venom. The county, along with private individuals, paid her to remove renegade snakes, scorpions and spiders. But the majority of her income came from selling antivenin to hospitals and clinics around the United States, Mexico and Central America.

After feeding the boas (live rabbits - bred for the purpose by Thelma), I spent some time cleaning tanks and hanging out with an albino Burmese python – a pink-eyed beauty by the name of Snowdrop - who had the run of the house.

After Thelma brought me my third cup of coffee, I had to ask if I could use her bathroom. She gave me a one of her odd looks and said, "Sure. It's at the end of the back hall. Just past the dining room. But be careful not to close the door when you're done. Arthur likes to sleep in the bathtub."

"Arthur?"

"Uh-huh. He'll either be in there or the dining room at this time of day."

As it turned out, remembering to leave the door open wasn't a problem; I wasn't even able to close it. Arthur, a twenty-one feet long, three-hundred-pound reticulated python, was neither in the bathroom nor the dining room. He was in both. His head basked in the rectangle of sunlight coming through the dining room window, while his tail cooled off in the ceramic claw-foot tub.

That afternoon, I accompanied Thelma to a builders' supply yard where a Western diamondback had been terrorizing customers and employees throughout the morning. When we arrived, a secretary, who was smoking a cigarette in the parking lot, told us the rattler had retreated to the crawl space beneath the modular building that housed the company's offices. Thelma kept a wide range of equipment for capturing snakes in the back of her van, but on this occasion the only preparations she made was to clip a flashlight to her belt and pull out a carrying crate.

A wide semi-circle of workers had formed around the torn off plywood opening where the snake had last been seen. None of them noticed our arrival. Thelma made a clicking sound in the back of her throat and then dropped the crate on the ground some distance behind the crowd.

"Open it up, Jason. And enjoy the show." She winked, then wandered over and tapped the largest man in the group on the back. "Say, is it true a snake went under this building, sir?"

"Yes, ma'am, it did." The big man peered down at her and touched a finger to the brim of his cowboy hat. The hat didn't move. It was held in place by pudgy temples and Velcro buzz cut. "That's why we're here. Don't want it getting out and hurting nobody."

"Yes, hurting nobody would be a terrible thing, I'm sure. Aren't you scared, sir?"

The man smiled and eased back his jacket to reveal a revolver wedged between his paunch and his jeans. "Take more than a rattler to scare me, lady."

"Uh-huh, I see. So why haven't you gone in to get it? Big man like you, with a gun, too. You should be able to handle a dainty little diamondback. Unless you're bullshitting me, and you're as scared as you should be after riling up the poor creature."

"What's your problem, lady?" He looked Thelma up and down. "You are a lady, right?"

"My problem is you and all these other macho pussies." She waved her hand toward the other men, who were now all listening in on her conversation with the big guy. "You're all too busy trying to look tough to properly understand and respect a fine animal that's done you no harm, in spite of all your provoking. You've got it corralled in and wound up tight. Don't think I can't see the scuffs your bullets made in the dirt there. You fools have made my job ten times more difficult than it should be. And to answer your other question, yes, I am a lady, you un-chivalrous bastard. But I'm also more of a man than you'll ever be, so I understand your confusion."

Thelma pushed past the dumbstruck cowboy and stooped to enter the gap in the plywood board, unclipping the flashlight from her belt and igniting it like a Jedi as she went. A chorus of whoops, a couple cries of "be careful", and quite a few more "What the hells!"

accompanied her disappearance into the gloom. Then the diamondback's echoing rattle silenced everyone in an instant.

Thelma's flashlight swept in a wide arc, then dropped to the floor as she jerked away from the snake's attack. Quickly recovered, she seized the snake and rolled out into the sunlight, gripping its thrashing body between elbows and knees. With a dramatic grunt, Thelma jumped to her feet, the conquered serpent, cowed yet still rattling, in her hands. She walked through the crowd, parting it like the red sea, and lowered the snake into the crate beside me. She snapped the sliding hatch shut and gave me another wink before turning to face the applause.

She curtsied, and in an exaggerated Southern Belle accent, said, "Which one of you strong men will help take this itty-bitty crate to my van for me?" Then, in her normal voice, she added, "And who do I see for my paycheck? I would ask one of you to retrieve my flashlight from under the building but I believe you're all a little too fat and scared." She turned to me. "Could you get it for me please, Jason?"

I grinned. "Sure, Thelma."

As I strolled through the crowd, I noticed the big cowboy nod to one of the men in overalls, who then rushed over to pick up the crate. The big man took off his hat, said something to Thelma, and then led her up to his office to receive her check.

Thelma was my new hero. I spent the rest of summer learning all I could about serpents, reptiles, arachnids and insects from her. I'd heard of herpetologists before, and for a long time I entertained thoughts of training to become one. But strangely, as much as I enjoyed my time with her, working with Thelma changed all that. I was like a schoolyard scrapper training with the heavyweight champion of the world. I knew I'd never be in her league.

Thelma's level of dedication was beyond anything I thought I was capable of achieving. She was also much smarter and braver than me. Of course, it helped that she was stark raving crazy too. I loved walking in her shadow though, and worked with her as much and for as long as I could.

Like all good things, my apprenticeship with Thelma came to an end. I'd decided to return to Alaska for tenth grade. I missed my dad and wanted to get away from Adrian. I'd also stayed in touch with Daniel, my daredevil rabbit-hunter friend, and he, along with my father, had spent the entire preceding year trying to convince me to return. Tales of sea-fishing, deer and moose hunts, and, most compelling of all, gold prospecting had piqued my interest and helped make up my mind to go.

*

A week before I was due to leave, the body of a young woman was found in the woods by our house. The newsreader said she'd been discovered, naked and horribly mutilated, on a flat rock outside one of the old cabins. I saw the footage of the site and recognized it at once. It was where we'd found the box of horseflesh.

Sheriff Bob appeared on the evening news, appealing for anyone who might have any useful information to come forward. "Even the most insignificant detail could prove pivotal to the investigation," he said, mustache twitching. I figured he probably remembered the horseflesh, and the horseshit stories we told him about bloodstains and pentagrams, so I didn't get in touch to remind him.

The woods were out of bounds while the police conducted their investigation. Apparently, the murder weapon and motive had yet to be discovered. The victim was a seventeen-year-old runaway drug-addict from Houston.

The autopsy reports were leaked to the local newspaper. They said a very sharp blade had been used to remove her heart and other organs, possibly while she was still alive. Everyone speculated about what had happened, with most suspecting it had been some kind of ritualistic, sacrificial killing. A local real estate company capitalized on the tragedy by petitioning the county to give over the land for development, extolling the virtues of safe housing over the threat of wild woodlands. They got the land.

The day the police allowed people back into the woods was the day I was due to fly to Anchorage. As my flight wasn't until the afternoon, I decided to go say goodbye to the grandmother tree in the morning. It was a dreary day. Drizzly, with a cold wind blowing in from the north. Appropriate for such a somber occasion.

As usual, I stood on the seat of my bike to reach the lowest branch and then took off up into the tree. As I climbed, my mind wandered. I considered the hunger for power, the ruthless cruelty, which I imagined led to the innocent girl being murdered. I thought of animals and plants, trees and birds, and wondered what they must make of human beings and our lust for blood and subjugation. I contemplated the damp glistening branches of the grandmother tree as I skipped ever upward. How long had she stood guard over this forest? How many life and death struggles had she witnessed? How many wasteful human crimes?

It struck me that trees are one of the most generous and underappreciated entities on the planet. They give fruit, seeds, shade, shelter, fuel. They even exchange our poisonous exhalations for breathable oxygen. In return, they demand nothing but to be left alone to continue to grow and give.

I was now seventy feet up and the branches were thinning out. I could see the dark sodden canopy stretching out all around me. To the north, the sun was trying to push through the rain clouds. Patches of gold and silver gleamed at their misty edges, forming a faint undulating rainbow.

I pulled myself further up the tree, my feet automatically finding the branch that had been my handhold a second before. My attention was fixed on the shifting dance of hazy color made by sun and evaporating moisture, until my reverie was shattered by a sharp crack and I slipped downward.

The branch where I'd been standing had snapped beneath me. If it had been a dry day I would probably have recovered with ease, but it wasn't; it was wet and slick and treacherous and I fell. Seventy feet up reduced to sixty in a split-second of thrashing arms, then my face hit a thick branch, spinning me into a blur of wet leaves and lashing twigs. Fifty feet. Forty-five. My world had been reduced to a blur of sopping, slapping greens and browns.

Although I knew I was in serious trouble, my mind grew oddly calm and detached. My soul was at ease with the prospect of imminent death. I wasn't brave enough to take my own life, but had no problems letting it go now the decision was out of my hands. After all, maybe Gwen was waiting for me on the other side.

Time stretched so taut it almost stopped, and I heard a woman's voice calling to me in a rasping whisper, urging me to open my eyes. Up until that point, I hadn't realized they were closed. When I did, I focused, through twenty feet of dense foliage, on a fork in the trunk below.

The moment I made up my mind to reach it, the branches in between started moving in ways that helped ease my fall. Some swayed away, others loomed in to catch my attention, so I could grab them with my hands or kick at them with my feet. All at once, I was there at the fork, and I shoved my body into the notched gap, my chest slamming into the hard wood. Gasping for breath, I braced my arms against the duel tree trunks so I wouldn't slide out. Legs dangling and head thumping I gave a whoop of relief.

My first thought was that trees might be a lot more generous than even I'd suspected; this old friend had moved her branches to save my life. My second thought was more grounded, however. Trying to make sense of what'd happened, and what I was now thinking, I decided I was probably suffering from a concussion. Needless to say, I took the last twenty feet of the descent a lot more carefully and slower than the previous fifty.

When I reached the ground, I said farewell to the grandmother tree. Still in a daze, I thanked her for the brutal lesson in focus and respect.

By the end of the day, I was back in Alaska, and by the end of the following year I would get my first taste of the obsession that would change - and almost destroy - my life. The fall from the tree, and the lessons it taught me, already a distant memory.

Jason Pednault   Matheu DeSilva

Gold Rushing

"Bears!" Daniel insisted. "I'm telling you, they're smarter than people think."

"Maybe so, but they're not as smart as people."

We'd been discussing *Douglas Adams'* assertion that certain animals - dolphins, for example - were potentially more intelligent than human beings, and that we only haven't noticed because of our inherent arrogance and narrow-mindedness. Daniel had made a compelling case for whales and elephants, and was now championing the cause of the king of the local fauna.

"Uh-huh." Daniel leaned back into the sofa, hands linked behind his head. "What about the fishermen of bird creek? They're supposed to be smart people, and bears train those fools to catch fish for them."

"What? How?"

"Don't you read the newspapers?"

I shook my head. "Only the funnies."

"Tell you what, let's go at the weekend and you can see for yourself."

Friday night we got our supplies together, including a shotgun and some bear-bells, just in case we got a little too close to the creatures we were hoping to observe. Saturday morning, a couple hours before dawn, my dad drove us out to Bird Creek. He went back into the city to work while Daniel and I hiked up to the ridge overlooking the river.

I'd been back in Alaska for nearly a year, and things weren't so bad this time around. Probably because I was older and able to get out more. During the winter months I'd spent my time snowshoeing, skiing, three-wheeling and ice tunneling with Daniel. Now it was summer, the season of hiking, fishing and long evenings. Which for us meant sneaking six-packs, avoiding parents at home, and moose and bears while outside.

We'd seen quite a few of both out in the wilderness already this year, and a couple each in the city, too. All from afar. Which was fine by me. This was the first time we'd actively gone looking for bears though, and as we climbed the switchback trail in the pre-dawn darkness my heart beat like a kettledrum.

"It's a good job I'm not epileptic," observed Daniel, who was walking a few paces in front of me. "The way you're darting that flashlight back and forth would give me a seizure. It's like a fricking disco strobe light."

"You're leading the way. I'm checking for bears."

Daniel laughed. "Jason, it's salmon season. The bears are all down by the river. Besides, Griz is on bear duty, aren't you boy?"

At the sound of his name, Daniel's massive dog came crashing out of the brush. He almost knocked me over in his enthusiasm to push past and reach his master. I wasn't sure I trusted Griz to keep us safe. Not only was he old and cantankerous, but he also appeared to have a suspiciously high amount of ursine blood running though his veins. Hence the name Griz. When push came to shove, or razor-clawed swipe, I wasn't sure where the bear-dog's allegiance would be.

The sky was lightening in the east by the time we crested the ridge. We kept hiking until we found a decent vantage point above a bend in the river. An angler was setting up some sixty feet below. Daniel tapped my arm and nodded further upstream. Our first bear sighting of the day; an enormous grizzly, out in the flow of the river, fishing serenely and efficiently with its spiked paws.

We watched the bear devour its breakfast in the distance while munching on the peanut butter and jelly sandwiches Mary had made for ours. When it was done, the bear stood up, sniffed the air, and peered downstream. It then lumbered further out into the river and repeated the action. This time it had a clear view of the fisherman, who was too engrossed in pouring coffee to notice the attention he was receiving.

Seconds later, the man's line sprung taut and he ditched his drink to concentrate on hauling in a fat silver salmon. We watched him land his catch and then attach it to a loop of wire he was wearing over his shoulder and across his chest, bandolier style. He reset his hook and cast off again. I glanced upstream. The bear was gone.

"Where did it go?" I asked Daniel.

"Into the woods. Keep watching."

A couple of hours passed, during which time the angler caught and hung so many fish at his waist he began to look like a particularly unattractive, bearded mermaid.

"I wish I was down there. The fishing here is great."

Daniel chuckled. "Yeah, it is. That's why the idiot got here at dawn like us. Figured he'd have the river to himself. Another one who didn't read the newspapers."

"So what's been happening?"

Daniel frowned. "You're asking me that now? I thought you wanted to see for yourself. What's the point in us coming all the way out here if you just wanted to be told about -"

"Shh, look!"

Daniel grinned. "This is it, keep watching!"

The bear we'd seen earlier emerged from the forest behind the fisherman. Griz's ears pricked up but he didn't make a sound. Like us, he was watching the bear approach the fisherman far below, his expression inscrutable. I reached for Daniel's shotgun.

"No." Daniel put his hand on the gun and shook his head. "Just watch."

The grizzly was ten feet from the fisherman now. The man still hadn't noticed him. My heart was in my mouth. I was certain we were about to witness something gruesome and couldn't believe Daniel was being so nonchalant about it.

The bear made an interrogative noise that echoed in off the cliff, and the angler nearly jumped out of his skin. Daniel cracked up. Even I had to smile. The startled man stumbled back and fell on his ass with a splash. The bear reared up and peered down at the terrified fisherman, who was now scooting back on his rear end, trying to get as far from the towering grizzly as possible without drowning.

The bear sniffed and gave the guy a meaningful look. Comprehension, or inspiration, must have hit him because he pulled the wire over his head and tossed his morning's haul at the bear. The bear pounced on the neat ring of fish, scooped them up to its chest and sauntered back to the shore where it gobbled them up. All the while keeping a gimlet eye on the hapless fisherman, lest he should move or show any sign of protest.

Second breakfast devoured, the bear wandered back into the woods. Presumably to do what it is bears are famous for doing there. The early morning angler waited fifteen minutes in the freezing water before daring to move. When he did, he hustled up to the beach, snatched up his things and jogged away down the shore to his car.

"So how smart was that guy?" said Daniel once we'd stopped laughing.

"Not very. He should have had a dog or a gun, or both. And why did he hang the fish from his shoulder? He'll be lucky not to attract more bears now. He must stink of fish."

"Nah, I reckon the smell of the fish got washed away and he only smells of the shit in his britches now. Anyway, I wasn't talking about the fisherman. I was talking about the bear. See how he let that dude catch a whole bunch of fish before coming out to collect them? He'd been watching him for a couple of hours. Biding his time. You saw that right?"

"Right," I agreed.

"So who's smarter? Man or bear?"

"When it comes to food, the bear." I conceded.

"What else is there? Food, shelter, taking care of yourself and your family - that's all that really matters. I'm sure bears think we're dumb for overcomplicating stuff. Think about it, we evolved all our hair off just so we can buy clothes! That can't make sense to any rational mammal."

I hummed the tune to *The Bare Necessities* from *The Jungle Book*.

"Yeah man! Old Baloo was right!"

"Uh-huh, Yogi Bear, too. But they happen to be smarter than the average bear, Boo-Boo."

Daniel rolled his eyes. "They're all smart. Admit it. We humans think we got it all figured out, but we can't even take care of our families right."

I thought of Daniel's dad, old Jimmy Roach, the violent alcoholic. And of his mom, who ran off when Daniel was a baby. I nodded and said nothing more.

We watched the grizzly do the same thing on three separate sets of fishermen. Only one of them had a gun; an old-timer, fishing alone. He let off a round when the bear closed in, but the bear stood his ground. The guy either didn't have the heart or was too afraid to shoot the bear, so he ended up capitulating and giving up his catch like all the rest. Both the other groups had dogs with them. They gave their owners fair warning but no real help. The bear made it clear he'd happily settle for a yipping terrier if no fish were forthcoming, so, of course, he received his tribute from the dog owners too.

Daniel and I decided to head back once our own food and water had run out. It was mid-afternoon and we figured if we walked slowly we would meet my dad at the trailhead just as he arrived to pick us up after work. We spent the hike back along the ridge laughing and imitating the various ways the folk had reacted to the bear. By the time we started our descent through the trees, we were all laughed out and fell into a companionable silence. Not a good idea in bear country, as we were soon to discover.

We rounded a switchback and came face to face with a startled mama grizzly. She was smaller than the one working the fishermen, but plenty big enough to handle Daniel and me. Griz barked and backed up behind Daniel, who was already behind me. The bear was thirty feet away, stamping and pawing the ground, puffing, and moving her head side to side.

"Good bear," Daniel said, grabbing my belt and pulling me back. "We're sorry to have disturbed you and we're going to go now. Nice and slow. Don't worry about us. We don't want anything. And we haven't got anything either. You go about your day and…oh shit! Hold your ground Jason! Hold your ground!"

The bear charged at us. Daniel was gripping my shoulder and belt, holding me in front of him like a shield. I was petrified. The reptilian part of my brain was screaming "Run!" but my body didn't respond. Somewhere in my frontal cortex another voice was urging me to keep still; running was the worst thing to do in any kind of bear encounter. Knowing that fact was no comfort. Not with four-hundred pounds of angry grizzly mama thundering toward me.

"Don't kill me! Please don't kill me!" I yelled.

Still she kept coming. Griz was barking up a storm and Daniel thrust the bells into my hand. I clattered them together, trying to make as much noise as I could. A metal tube struck my temple, making me stagger sideways.

The bear was almost on top of us, and I fought the urge to throw myself into the undergrowth and let Daniel and Griz take care of it - or be taken care of by it. A loud crack-boom filled the air, freezing the moment in a tableau of bared teeth, terror, claws and ringing ears.

The bear stopped and backed away, huffing and stomping, still upset. Daniel shook my shoulder and mouthed something about the bells. He pointed to the bear. Eyes wide. He was shouting, but the noise coming from his mouth was distant and muffled. I came to my senses and banged the bells together over and over again - though all I could hear was the high-pitched ringing Daniel's gunshot had left in my ears. I shouted gibberish and stamped my feet too. Daniel raised the gun a second time, and, still shouting, I put my fingers in my ears. He fired another warning shot.

"Go bear," he screamed, reloading the rifle with mechanical

proficiency. "I don't want to kill you, but the next shot's gonna blow your head off if you don't leave us alone."

The bear backed up a couple more paces. She seemed intent on charging us again until something in the undergrowth, down and away to our left, caught her attention. Scampering around the hill were two fuzzy little cubs. Mama bear glanced from us to her fleeing young and back again. She made as if to charge one more time and then veered off through the vegetation after her children.

"I told you they were good parents," said Daniel, his voice shaky. He lowered the rifle. "Let's get out of here. I don't want to be here if she comes back."

My ears were still ringing, but my hearing was slowly coming back. Along with a sense of relieved equilibrium. "Me neither. Let's go," I agreed.

Not wanting to surprise any more bears, we ran the rest of the way to the trailhead wailing like banshees. Griz barked and lurched at butterflies as we went, as if trying to reassert his authority over the wild. Breathless and wired, we arrived at the parking lot to find my dad sitting in his van reading a newspaper. He took a long look at us through the window before leaning over and unlocking the passenger door.

"What's with all the noise?" he asked, as we bundled in.

"Bears," I managed to say between gulping breaths.

"Oh." He put the van into drive and pulled away without another word.

Bears were, and still are, so ubiquitous in Alaskan life that encounters such as ours would scarcely raise an eyebrow up there. Only a serious mauling is worthy to be called a real bear story. However, my father did enjoy a good chuckle at the expense of the naïve fishermen, whose misadventures with the big grizzly we described at length on the drive back.

Over the intervening years Bird Creek's popularity with fishermen has grown exponentially. Cheek by jowl 'combat-fishing' is now the norm at high season. Put off by the large number of people, bears stayed away for the most part.

In recent years, though, they've made a bit of a comeback. It seems grizzlies are once again letting the foolish humans do their fishing for them. When I saw a news article about this returning phenomenon, I couldn't help wondering if those scampering fur balls whose mother we met on the trail back from the ridge, are now part of the resurgence of their father's trade.

On our way home, we saw Daniel's father hustling down the street toward the local bar. As soon as I spotted him, I asked Daniel if I could come over to his place in the evening.

He laughed. "I'm not sure she's going to be there, Jason. Maybe she'll be out on a date or something."

My dad glanced at me, the corner of his mouth twitching. My ears burned. We all knew who Daniel was referring to.

Aside from anything else, that was the summer of the half-sisters. Both my half-sisters from my dad's first marriage - Monique and Chantelle - had come to live in Anchorage, giving me the opportunity to get to know them for the first time. But the half-sister who really caught my attention was Daniel's.

I'd always assumed Daniel was an only child. He'd never mentioned his sister until the day she came up from Missouri to live with him and his dad. Sammy was two years older than Daniel, and a year older than me. A high-school drop-out, she worked in one of the two fashion boutiques downtown. She was trashy and gorgeous, and only dated older guys who had nice cars and bad reputations.

While up on the ridge, Daniel had told me she'd broken up with her latest boyfriend and had been moping around the house the past few days. I knew Sammy was way out of my league, but even the lowliest underdog has dreams of one day making the play-offs.

I went home, had a shower, gelled my unruly hair, splashed my dad's *Old Spice* over my neck and torso, pulled on my best shirt, scarfed my dinner, appropriated a six-pack from Dad's stash in the garage, and set off for Daniel's house. On the way, I popped open one of the beers and chugged it down. Seeing as I considered this my best chance of finally getting Sammy to notice me, I didn't want to ruin it by being sober.

Daniel opened the door. He coughed and grinned. "I smell, sorry, I *see* you've made an effort. Sammy's pouting and painting her nails in the backyard. You want to go through and see her or play some video games with me while you build up your courage a little more?"

I grimaced and handed him a beer. "Courage building sounds good. Let's do that."

After thirty minutes of shooting space ghosts and drinking beer, I'd almost forgotten Sammy was in the house. Then she walked into the living room and my stomach tied itself in a knot so tight, I felt the effect all the way up to my strangled tongue.

"Hey Jason. What you doing?" she purred, or slurred. I couldn't tell which, but hoped it was the latter.

I stared at the screen, where my avatar was being shot and killed repeatedly. My stiff and sweating fingers could do nothing whatsoever to save him. Game Over.

I glanced in her direction. "Hey Sammy. You say something?"

She rolled her eyes and collapsed into the armchair. "Nope. Nothing. Boys and theirs games, huh? What's a girl to do?"

What a question. Nerves and beer were fuzzing my mind, but I scrabbled to answer her. "I don't know. What's a boy to do?"

She laughed. It wasn't the tinkling delicate sound she usually made but a raucous belly laugh that set me off too. Daniel joined in, and the three of us giggled like little kids. I'm guessing only one of us knew why.

"You're too much," she said, wiping her eyes with the palm of her hand. Scarlet nail polish. Pale slender fingers.

"So are you," I blurted, which started her laughing again.

I felt like the wittiest man alive. Then I spied the half empty bottle of *Everclear* dangling from her right fingertips. Knowing she was more drunk than me was a relief; it helped me grow in confidence. It didn't matter it was just the drink talking and making her laugh - we were connecting, damn it!

She caught me looking at the bottle and hid it behind her. "You want some?" she asked, arching her perfect eyebrows.

"Sure." I swaggered over to the armchair, pried the bottle from its snug little hiding place in the small of her back and raised it to my mouth.

I'd never drunk strong liquor before and the fumes coming off the bottle made my eyes water. *Everclear* is a grain alcohol. It's tasteless and very strong. It's the perfect drink for a no-thrills alcoholic. No wonder their dad loved the stuff. I took a tiny sip. My throat, nostrils, and entire head burned to the point of numbness.

Sammy giggled. "Don't kiss it Jason. Drink it! Like a man! You are a man, right?"

I nodded and took a huge swig. Enough to make me cough and wheeze like a very old man. Bile rose in the back of my throat, but I managed to swallow it back. Sammy writhed around in the armchair, cackling hysterically.

Daniel took the bottle from me and drank. He grimaced but didn't replicate my excruciating display. "You do know Dad's going to be furious when he finds out we're drinking his liquor?" he said to Sammy.

"That's the best part! The old bastard always comes home so drunk that come morning he'll think he drank it himself."

We all found her observation hilarious. We fell about the floor laughing, and continued to drink the *Everclear*. At some point, Sammy turned off Daniel's *Atari* and put a *David Bowie* cassette into a tinny pink portable sound system she'd dragged in from her room. We danced around like lunatics, striking poses, thrashing our arms about and bumping into one another. Then a slow song came on, and Daniel fell back into the armchair, tactfully closing his eyes. Sammy and I gazed down at him.

"I feel terrible," he said, crossing his arms over his eyes for good measure. "Damn, I can't believe Dad drinks this stuff for fun."

"But it is fun," said Sammy, pulling me toward her.

We danced close and slow, and at the end of the song she kissed me. Not the chaste sort of kiss I'd shared with Gwen. This was a full-on adult kiss. Clinking teeth and hot seeking tongues. I'd never known anything like it. My fantasy was coming true. Sammy pressed her body against mine, guided my hands all over her. My head swam and my vision grew grainy. Then everything went black.

When I awoke, my arms were no longer around Sammy. I was embracing something cold and solid, which, when it came into focus, I recognized as the porcelain toilet bowl in the bathroom off of Daniel's living room.

The sight and stench of vomit made my head pound. I reached up and flushed. The water rushed round the bowl and the cistern filled, roaring and hissing; an exquisite torture for my tender brain. I felt as though someone had driven a freight train through my skull, stopping along the way to take a crap in my mouth. My throat was raw and dry and my arms were stiff and aching. I clambered to my feet and checked the mirror. Bloodshot eyes stared wildly back at me from above my vomit speckled nose and chin. Voices were being raised outside the bathroom. I watched my face crinkle into a frown as I strained to hear what was being said.

"What the hell you think you was doing getting your sister fall-down drunk on my liquor you little bastard?"

Jimmy Roach was home.

There was a noise like meat hitting a chopping board, and I heard Daniel groan and cough. Without thinking, I burst through the bathroom door. Two pairs of pink rimmed, pale blue eyes whipped around toward me. Jimmy let go of Daniel, who fell to his knees.

"What you doing in my house you little shit? Did he let you in here?" He pointed at Daniel. "Trying to take advantage of my whore of a daughter, were you? Bet you both tried to fuck her. Sick little bastards." Jimmy's eyes were bright with glazed fervor. Spittle hung from his bottom lip.

I glanced around the room. Sammy was gone. Daniel was cowering on the floor. He mouthed *get out of here* To me.

I couldn't leave my best friend alone with this maniac, even if old Jimmy did scare me more than the grizzly we'd faced that morning. There were guidelines on how to handle bear attacks. Violent, raging alcoholics were as unpredictable as high seas in a hurricane.

"Answer me boy!" yelled Jimmy.

I panicked, realizing I couldn't remember his question. Jimmy lunged at me, swinging a haymaker with his right fist. I was nauseous and terrified, but my reflexes were still good. I ducked and slipped under the punch. Jimmy's momentum carried him forward and he staggered, lost balance and fell, smashing his forehead against the countertop of their home-bar on his way down.

Sammy's familiar tinkling laughter came from the doorway across the room. She sounded unhinged. Mascara streaked in tear tracks down her cheeks. I wanted to run to her, but saw she was holding her father's gun. Hands shaking, she raised it and aimed it at me.

"Sammy, what are you doing?" Daniel called out from the floor. It was the question I was too stunned and afraid to ask.

Eyes still locked onto mine, she swung the barrel around and pointed it, one handed, at her father, who still lay unconscious on the floor. "I'm going to kill him," she said, her eyes welling with unshed tears. "The old bastard deserves to die."

I nodded. "He does. But you don't deserve to have his life on your conscience." I had no idea where that piece of wisdom came from, but I hoped it might stop her from blowing her father's brains out in front of me.

"Get out!" she screamed, glaring at me. The end of the pistol dipped and wavered. "You weak, sick little boy. Get out of here!"

Daniel appeared at his sister's side and plucked the gun from her trembling hand. Without the weapon holding her up she crumpled to her knees and broke down in heart-wrenching sobs. Daniel put his arm around her, then coaxed her to her feet and led her away down the hall.

A head spinning stillness settled over the room. Old Jimmy Roach snored softly from the floor next to the bar, a red lump the size and shape of a boiled egg throbbing in the center of his forehead. I left and staggered home, stopping every now and then to throw up along the way.

*

When I saw Daniel a few days later he told me Sammy had gone back to Missouri. She'd told him to tell me she was sorry, but didn't specify what about. I never saw her again. He also told me his dad hadn't remembered anything of what had happened. Apparently, he'd awoken next to his bar with a splitting headache and the near empty bottle of *Everclear* close by. He'd put two and two together and, just as Sammy had predicted, came up with five.

Jimmy was surprised and upset when Sammy said she was going back to live with her mother. She wouldn't give him a reason for her decision, so of course he blamed Daniel. It became another reason for Jimmy to beat up on Daniel after he'd had a few drinks. Not that he needed it. Jimmy Roach continued to stagger and punch his way through life, never knowing just how close he came to being shot dead by his own daughter.

Daniel once told me how his father had been a well-known and respected tracker and hunter in his younger days. He'd grown up with the native tribes in the north and had learned the outdoorsman skills he would later pass on to Daniel from them. He'd been a man with a purpose and a deep connection to the natural world. But somewhere along the way he'd found the bottle and lost it all.

I've seen many adventurous spirits nullified by booze or drugs over the years, and I'm sure if it wasn't for that night I might well have become one of them. Instead, I've avoided hard liquor as much as possible. For me, there's nothing but madness, sadness and disappointment to be found in the bottle. And life has more than enough of that without adding in the booze. Besides, my passion, and poison, was always meant to be gold.

*

A few weeks after Sammy left, my dad told me he was taking the weekend off so we could go camping. I was out with Daniel when he loaded up the van Friday evening, so it wasn't until I climbed in the following morning that I noticed the three-wheeler and all the extra equipment in the back.

"Where we going, Dad?" I held my coffee in one hand, and was trying to keep Chico, our overly-excited elderly cocker spaniel, from spilling it with the other.

Dad grinned. "We're off to make our fortune, Jason. This is the last chance we'll get to go sniping before the weather sets in."

"For real?"

He nodded, and it was my turn to grin.

I'd long since given up pestering him about his promise to take me gold-sniping. I knew he was busy, and I didn't want him to feel obligated when he had so much else going on. However, my father was a man of his word; if he said he was going to do something, nothing short of a full scale natural disaster would stop him following through.

We travelled inland, winding our way along service roads through forests of tall pines until all signs of civilization were behind us. By midday we'd travelled as far as the van could take us. We then transferred our kit to the three-wheeler and continued, with Chico running dutifully alongside. A couple hours later we stopped and hid the three-wheeler in some brush, packed everything into our backpacks and began hiking.

Mid-to-late afternoon we made camp at a small clearing next to a fast-flowing river littered with boulders. After eating lunch, we hoisted our food up into a tree away from bears, then got down to the serious business of gold prospecting.

The idea of gold-sniping is to find silt filled cracks and crevices in boulders, clean away the top layer of debris, and rummage around for the gold that has sunk, over time, deeper inside. Often small (or, if you're lucky, quite large) nuggets and clusters of flakes can be found this way, making it a pretty efficient way to prospect.

The site my father had chosen was perfect. All the boulders were crisscrossed with cracks and most were only a foot or so beneath the surface of the water. Larger rocks were strewn across the width of the river, poking up through the surface like stepping-stones made for giants. We leapt from one to another, making our way to a place Dad thought would provide a good yield of gold. Chico jumped and scrabbled along behind us, not wanting to miss out on the adventure.

We sat on a low flat rock and peered over the edge at the boulders beneath the surface. My dad used a hand pump to remove the top layer of sandy debris, which he deposited into a bucket he'd brought along with us.

"We can pan that later," he explained.

I gazed into the bucket. Here and there a flicker of shining yellow caught my eye. I was ready to start panning immediately; gold fever was already setting in.

"Check that out, Jason." My dad was lying on his belly pointing at the boulder he'd been cleaning.

I leaned over to see. Nestled in a wide crack was a lump of the same shining yellow I'd glimpsed slivers of in the bucket. "Is that gold?" I asked stupidly.

"Sure is. You want to get it, or should I?" He pulled a pair of slim long-nose pliers from his pocket and held them out to me.

The water was clear but swift. I wanted to grab the gold but felt nervous. If I didn't get a good hold on the nugget it would tumble away downstream and be lost forever. I wasn't prepared to let that happen.

"You," I said. "I want to see how it's done first."

My dad nodded and shimmied further out, closer to the boulder. His knife hand dipped under the surface and a look of deep concentration fell over his face as he pried around the gold nugget. "Grab my legs, Jason. Looks like this sucker's wedged in deeper than I thought." He smiled and winked at me. "Which means it's bigger than I thought, too."

I sat on the back of my dad's thighs and peered over his shoulder while he plunged his arms into the water. He ran the tip of the pliers across the boulder against the current toward the nugget. He paused on the edge of the crack. Opened the pliers. Reached in and gripped the nugget.

"You got it!" I said, slapping him on the back.

"Easy, Jason," he hissed.

I didn't move or make another sound. In fact, I held my breath as he used his knife to help ease the piece of gold from the crack. Once out, he drew it through the surging water and into the air. Sunlight glinted off its irregular surface. It was beautiful, if somewhat smaller than it had appeared when magnified beneath the water. He held up the pliers for me to get a good look at our prize then plopped it into a small black cylindrical container that had once housed film for his Nikon camera. He handed me the container and I peered inside, still mesmerized.

"How much is it worth?" I asked.

"Hmm, about seven, eight bucks, maybe even ten. We'll have to weigh it to be sure."

My mouth fell open and I stared at the little lump of raw gold. I worked part time at the local pet store, taking care of the snakes and lizards, for four dollars an hour. My dad had just plucked eight dollars out of a rock in a river. It took him seconds to make what I made in two hours of scrubbing the crap out of glass tanks.

I sat back and looked around. Our camp was made on a grassy glade at the water's edge. Above were high snowy mountains, stark and eternal. Below, the river fell away into a wide wooded valley. The air was fresh and the water pure. Fish and animals were abundant, as were wild herbs and berries, and all the money anyone would need was sitting in the rocks just waiting to be harvested. I was confused.

"Why do you install flooring, Dad? Why don't you look for gold full-time instead? We could live out here in the forest if you did."

"Good question." He didn't look up. He was busy scraping away at the crack that had yielded our nugget. "Responsibilities, I guess. It's good to know you have a steady wage coming in. Besides, as beautiful as it is out here, the weather can be unpredictable and the water's freezing. Speaking of which, it's your turn to get wet. You want to pick that one out for us?"

He pointed the tip of his knife at another nugget. It wasn't quite as large as the first, but it wasn't far off. My dad held my legs as I reached into the icy river. The current tugged at my arm, and I had to fight to keep my hand steady. Employing the same technique as my dad, I sidled the pliers up to the crack, seized the nugget and used the knife to help pry it out.

I did it. From that moment on, I was hooked. I knew exactly what I wanted to do with my life: I wanted to live in the wilderness and hunt for gold.

Over the next few hours my dad taught me how to use the pump and the scraping knife, and, most importantly, how to work out the most likely places to find gold in the river. He also showed me how to pan the fine sand and get the gold flakes out.

"One more crevice and we'll get started on dinner, eh?"

At the word dinner, Chico, who'd been basking on the flat rock beside us all afternoon, sat up and waged his tail. My dad's suggestion had the opposite effect on me. All I cared about was finding more gold.

"Can't we keep going? There's at least four hours of daylight left."

"We've got all day tomorrow, Jason. Besides, I'm hungry, and you can't eat gold. One more and we're done for the day, okay?"

I gave him a sullen nod and set about cleaning the final crevice. Chico was dashing back and forth across the rock behind me, no doubt wondering why we were still dawdling and messing about in the river. I turned to deposit the silt from the pump into our bucket, and Chico rushed forward to see what I had.

His paw struck the film case that held all our gold, knocking it over and setting it rolling down the curve of the rock. Both my dad and I made a grab for it at the same time. Our heads cracked together, and the case slipped between our grappling fingers. It tumbled away, with Chico now in pursuit. I had a vision of him snatching up the case in his jaws and returning it to us, triumphant. What actually happened was the film case plopped into the river, closely followed by Chico. Both disappeared within seconds.

I couldn't help it, to my eternal shame, I had only one thing on my mind. "The gold!" I wailed.

"Never mind the damn gold," said my dad, already leaping back across the rocks toward shore. "We've got to find Chico!"

I snatched up our kit and followed him. We ran downstream along the river's edge, calling Chico's name and searching the banks for any sign of him.

"Maybe one of us should cross the river," I suggested. "He could have washed up on the other side."

"If he has he'll hear us. Better we stay together. Don't forget this is bear country."

After my close call with the she-bear above Bird Creek, the last thing I'd do was forget that. Still, once he mentioned it, I shouted even louder and kept closer to my dad, who had his rifle slung across his back. We searched for two hours before giving up for the night and trudging back to camp. We had no choice but to stop when we had. The sun was already low and would be gone by the time we reached our tent as it was.

"We'll try again in the morning," said my dad.

I nodded. The lost gold was forgotten. All that mattered now was Chico.

Neither of us slept well. It was too quiet and we were too miserable. Our two-man tent seemed huge without the faithful cocker spaniel wedged between us.

We rose early to renew our hunt for Chico. Our gold prospecting kit already packed away, we scoured the banks both sides of the river throughout the morning. Occasionally, I caught my father casting a furtive glance into an eddy. I knew he expected to see Chico's soggy brown and black fur churning in the water, buffeted by the trapped driftwood. We both did. But there was no sign of him anywhere.

"There's not much more we can do, Jason. Let's go back and break camp. We should go home."

"But we can't. We can't leave him out here."

I had tears in my eyes. So did my dad.

"He's gone, Jason. We have to face it and say our goodbyes to him here."

So we did. Dad talked a while about what a great dog he'd been, and how maybe this time around he'd get to be reincarnated as a husky. That made me smile, and then sob. Dad gave me a hug. I'm not sure, but I think he cried then as well. In spite of the bear danger, we returned to the camp in silence.

When we pushed through the brambles at the edge of the glade, we heard a dull rhythmic thudding coming from the side of our tent. "Ha!" said my father. "Chico, you little shit!"

I peeked over his shoulder and saw Chico lying on his belly beside the tent, bone dry and wagging his tail. The end of it was beating against the canvass, sending ripples out in all directions.

"Chico! Where were you?" I pushed past my dad and crouched to call the dog over.

Chico bounded over to us, and in a flurry of petting, mock scolding and dog licks, we were reunited. Dad squinted at the midday sun. "Let's get the kit out again. See if we can't find enough gold to at least cover our gas money." He turned to Chico. "*You* can stay on the shore this time."

We worked all afternoon and half the next morning, riding the high of finding Chico alive. We barely even stopped for food. In the end we went home with over half an ounce of gold, which in those days was worth around two hundred dollars, or fifty hours in the pet store, depending how you looked at it. To my disappointment, my dad never cashed the gold in, preferring to keep it as a souvenir, or a nest egg for a rainy day.

Even though that trip was supposed to be our last chance to go gold-sniping before winter set in, we still managed to squeeze in two more afterward. Both times we came back with more than half an ounce, and one time we had closer to three-quarters. All of which my dad stashed away.

Dad didn't really care about the value of the gold. It was the value of the experience, of spending time together in nature, of questing, which meant the most to him. Although I loved that aspect too, I still saw dollar signs in every nugget and flake.

Pretty soon I was going out prospecting by myself. Needless to say, everything I found on my solo trips was always converted into cold hard cash.

Jason Pednault   Matheu DeSilva

King of the Worms

It was early morning and I was up to my elbows in compost, sorting through a mass of worms, separating the smaller ones from the large breeders. The door to the greenhouse opened, and my landlady, Mrs. Wright, came in with two cups of coffee.

"I sure am going to miss you, Jason. But not as much as my garden will miss your wriggly little friends." She gazed out at her thriving veggie patch, then at the clumps of tomatoes weighing down the vines inside the greenhouse. "Do you know where you're going to from here?" she asked, placing my coffee on an upturned crate and handing me a rag to clean my hands.

"Back to Anchorage, to see my dad. After that, I don't know. Hopefully some place warm. Texas, maybe. I've still got family there."

I'd been working as an itinerant flooring installer in Alaska for five years, moving every few months to wherever there was work. Being Alaska, it was always somewhere cold. The short summer was coming to an end, and the thought of another long dismal winter in the forty-ninth state was making me even more depressed than usual.

"Why don't you spend a little time in the state park on your way back? You've got all that camping gear you hardly use, and you said you wanted to explore the wilderness around here more. May as well be now. You could even pan for gold like when you were younger."

I'd been lodging with Mrs. Wright since early spring, and at some point I must have told her about the gold prospecting adventures of my youth. "The park sounds good. I'm not sure about panning though. It's already a bit cold for that."

"If you say so. But if you keep thinking that way you'll never make your fortune. You want to work on what people walk on forever?"

I shrugged and picked up the bucket of baby worms I'd separated from my big tub. "These are for you. Something to remember me by."

Her face lit up. "Thank you, Jason! Are you sure?" She smiled and reached into the bucket to pull out a handful of the red wrigglers. "I'll never have to squirt chemical fertilizers on my food and flowers again," she said with a conspiratorial wink. "Who'd have thought these little guys could make so much difference. If more people knew what you knew, there'd be half the pollution there is in the world and a lot more productive and pretty gardens."

"Don't worry, I'm sure vermiculture will catch on once someone figures out how to make money from it. I doubt it'll be any time soon though."

She laughed. "You never know. Maybe someone could genetically modify a strain of wrigglers just enough to be able to patent them." I grimaced, and she laughed harder. "Okay, maybe not. Now, I don't have a parting gift for you, but if you do want to spend some time up at Chugach Park, I do have something I'd like to loan you."

I raised my eyebrows. Mrs. Wright was a fully domesticated townie. She would go into meltdown if her borders and hedges weren't trimmed to stifling neatness on a weekly basis. I couldn't imagine she owned anything that would be of use in the wilderness.

"Why don't you take Bucky?" she said. "He'll keep you safe from bears. And it'll give you guys a chance to say a proper goodbye to each other. I think he's grown even more fond of you than I have."

Bucky was Mrs. Wright's mastiff-German-shepherd cross. Apart from occasional lodgers like me, he was her only companion. They were devoted to each other and had an uncanny connection that bordered on telepathic.

When I arrived, she warned me Bucky never warmed to any of her paying guests. But after I'd spent a few weeks sneaking him bacon scraps under the breakfast table, he'd bound out to greet me when I finished work, and join me on walks to the edge of town during the long summer evenings. I was never quite sure how Mrs. Wright felt about the bond Bucky and I had formed, and was very touched and taken aback by her offer now.

"If you're sure, I'd love it if he came with me. I'll take good care of him."

She gave me an odd smile. "Funny. That's what he said about you."

*

By mid-morning I was ready to go. Worms and tools were stowed, and my camping gear had been pulled out and stuffed into a backpack just inside the van's back doors, along with food supplies for two weeks in the wilderness packed into daily ration kits, including hefty servings of dog biscuits for Bucky.

I couldn't remember the last time I'd been out in nature for an extended period and was both excited and nervous. I had butterflies in my stomach and a delicious aching that ran soul deep. It was as if I were going to meet an old flame I'd never fully gotten over.

When I pulled in to the gravel lot at the edge of the park, I just sat in the van and gazed up at the high glacial peaks, the broad sweep of forest. Rivers wound their way down through the clefts and valleys, glinting emissaries of the hidden lakes in the mountains beyond. I could have stayed and watched the shifting clouds and sun-dappled slopes for hours, but Bucky was having none of it. He was hopping back and forth and huffing on my cheek, eager to get out.

I heaved on my backpack and we hit the trail. Once the van was out of sight I felt a rush of manic elation. Pristine, rugged nature always hit me that way. Like an addict taking his first hit after months of self-imposed rehab, I couldn't understand why I'd denied myself the exhilaration of the high for so long.

Somehow, I'd fallen into the modern-day trap where making a living becomes more important than the actual living itself. I stopped and took a deep breath of the cool, clean air and resolved to escape the grind more often, if not permanently. Bucky ran ahead then circled around, bounding and frolicking like a pup. Seems I wasn't the only one feeling energized by the raw magic of nature.

The next twelve days passed in an idyllic blur of hiking and fishing. All the while I refused to think of the future or look back on the past. The ever-arriving moment was all that mattered. Sunlight passing through leaves, bathing in cold, still lake water, fishing for my supper, starry nights and sleeping beside the fire.

Then, on the thirteenth day, we found the cave.

We were walking a trail along the bottom of a wide, rocky outcropping when Bucky disappeared into a large bush butted up against the cliff-face. There was a rustle of leaves and then silence. I glanced at the bush, expecting to see him lying flat beneath it. But he wasn't there. Closer inspection revealed a five-foot by four-foot opening in the rock hidden behind the foliage.

I called into the dark recess, but Bucky didn't come out. I wasn't sure whether to wait for him or press on ahead. As I stood there trying to decide, heavy drops of rain spattered on my arms and shoulders. I looked up. Thick leaden clouds were rolling in from the north, obscuring the land beneath in a drifting vertical haze of rain. I glanced at the cave. Maybe Bucky had sensed the storm coming and had found us a shelter.

"Good dog," I said, already pushing through the bush toward the cave.

I ducked inside just in time. The rainstorm washed over the trail, turning the packed dust to mud within seconds. I expected to find Bucky curled up against the back wall of the cave, but neither dog nor back wall was there. The cave was bigger than I'd thought. Crouching, I clicked on my flashlight and ventured further inside.

After fifteen or twenty feet, there was a shift in the air; it had become cooler, more expansive. The cave had opened out into a large cavern. I straightened up and was immediately jabbed in the head by a sharp rock. Cursing, I stooped back down and cast my flashlight beam upward. Stalactites glistened and sparkled from the cave's ceiling, ascending away in waves, higher and higher. Another step and I was able to stand without hitting my head again.

A series of panting breaths echoed around the vast chamber, and Bucky padded toward me out of the gloom. "Clever boy," I said, patting his head. "Let's take a look at what you've found." I removed my backpack, laid it by the entrance and began exploring.

The main chamber was huge - at least forty feet across – with small tunnels and antechambers coming off at irregular intervals. As well as the thrill of discovery, this secret underground world gave me a reassuring feeling of womb-like security.

I had a nagging concern my flashlight might run out of batteries and leave me blind, so I turned it off and stood in the darkness, soaking in the peace and stillness. Bucky lay by my feet. His ease inside the cave made me feel even more comfortable and assured. If this were the lair of some dangerous beast, he wouldn't have been as calm as he was.

The rain continued to fall throughout the afternoon and into the evening, so I decided to make our last camp of the trip inside the cave. We made a foray outside for firewood. There was dry kindling beneath a giant pine not far from the cave entrance, and I was able to break off some seasoned branches from a bleached-white deciduous tree I'd seen lying beside the trail just before Bucky had disappeared.

Once the night's firewood was gathered and brought inside, I unpacked my sleeping mat and cooking kit and lit the fire.

Bucky and I ate a dinner of smoked trout and rice seasoned with wild herbs. Afterward, I lay back and watched the light of the flames flickering off the stalactites overhead. I knew I'd be back in Anchorage either the next day or the day after, and my resolve not to think of the future or past dissolved away in a flurry of deep, searching questions.

Why had I stopped doing the things I loved - gold prospecting, fishing, hiking, camping out in nature? Why wasn't I cultivating my own garden somewhere instead of hauling around worms to fix other peoples'? Why was I so miserable all the time? Why did I ignore this discontent and keep on trudging on? What the hell had happened to me?

I'd been gloomy and restless for as long as I could remember, but I'd never consciously thought about why, much less figure out a way to change the way I felt. I did what everyone else did. I drifted along. Like a corpse; face down in the river of life.

During my last year of high school, I'd spent all my free time either learning about gold prospecting and treasure hunting, or out doing it. I was obsessed. It lit a spark in me and gave me a passion and a purpose. But then, as is so often the case with childhood dreams, the expectations of family and society slowly beat that spark out of me.

By some miracle my grades were good enough for me to continue studying after high school, so, at my family's insistence, I entered Blinn Junior College. I wanted to major in Archaeology at Texas A & M. I was still naïve enough to think doing that would provide me with a way to travel and hunt for treasure, while still pleasing my parents and fitting into societal norms.

"Archaeology? What the hell's wrong with herpetology?" demanded Thelma when I told her my plans.

"You ever met a rich herpetologist?" I asked.

"More to life than being rich, Jason. You got to do what makes you happy. Money will come or it won't. And if you're happy in what you do, you won't care either way."

Arthur, the reticulated python, emerged over the back of her threadbare couch and settled his head upon her shoulder. The rest of his body was still making its slow progress through the hall and into the living room from the bathtub. Thelma reached up and tickled his chin.

We sat in silence for a while, until her eyes narrowed into snake-like slits. "Now I get it. I know what you're up to, you sneaky little bastard!"

"You do?"

"Sure I do. It's dinosaurs, right? You want to look for dinosaur bones!" She winked at me. "I knew you wouldn't let me down. Ha!

Now it all makes sense."

"I guess it does," I agreed, not wanting to disappoint her.

Dinosaurs did fascinate me, but I was more taken by the idea of ancient temples, forgotten cities and never-ending hauls of glittering treasure. Two and a half years later, I realized studying archaeology was more about ancient textbooks, forgotten hours and endless reams of test papers.

For every hour an archaeologist spends in the field, there are at least another ten-to-a-hundred squandered in stuffy rooms, cataloguing and writing up every little disturbance or anomaly in the soil in meticulous detail; hardly the life of adventure I'd craved.

I alleviated my boredom and misery by escaping into online multi-player fantasy-world games and immersing myself in the college party scene. Whether it was flirting with a dark sorceress in my paladin alter ego under the shadow of virtual dragon wings, or gurning and throwing shapes for five days straight after taking too much ecstasy at a warehouse rave, anything was preferable to the soul-sucking dullness of studying for my degree.

When the tutors noticed my lack of commitment they gave me an ultimatum: shape up or get thrown off the course. I decided to leave of my own accord and set up home with the sorceress.

Her name was Vicky, and away from the *Ultimate Online* universe where we'd met, she was a graphic art student from San Jose, California, with wine-red hair, china-doll complexion, a wicked sense of humor and a penchant for black, lacy clothes. Although she'd been upfront about her jealousy issues and pathological hatred of all things 'outdoors' from the beginning, it took seven years of suffering and denial, with a few good times thrown in here and there, for us to call it a day and go our separate ways.

That's not to say my time with Vicky was completely wasted. For one thing, I learned the family trade of flooring installation during that period. Also, in the early days, while I still had the passion, I picked up a lot more knowledge about gold mining techniques. Mostly from a friend called John who I'd met in Medford, Oregon, where we lived while Vicky studied for her graphics degree.

An engineer by trade, John had the gold bug as bad as I did. He bought a claim on a stretch of river near Klamath and asked if I'd be interested in working it with him. The gesture wasn't completely altruistic. He had a powerful gold dredge that he'd modified himself, and he needed someone to help operate it.

Dredges are essentially small barges with a pump, hose, and a series of sluices that make gold prospecting more efficient. The hose sucks up all the raw material from the riverbed and the sluices filter out the gravel so all that's left at the end is fine black sand and flakes of gold.

Most dredgers separate the gold from the sand using mercury. Gold binds with mercury. Sand doesn't. Once the gold has been absorbed out of the sand, the mercury is burned off, leaving a fat porous nugget of pure gold behind. Neither John nor I wanted to contaminate the riverbanks with mercury, so we used something called a *Blue Bowl*, which uses water and centrifugal force instead. It took a lot longer to extract the gold this way but it was far less polluting.

Dredging is hard work. A two-man dredge crew is made up of a hose-man and a rock-man. Both work under the water - or at least half-submerged - most of the time. Working the hose is like wrestling an anaconda. But as testing as that can be, it's preferable to being the rock-man, whose job it is to constantly sling big rocks out of the path of the hose end. If a sizable rock gets sucked into the hose it can stop work for a long time, so having a good rock-man is vital. John and I took turns to do each job. Often with Vicky scowling at us from the riverbank in stiletto-heeled boots and a flowing black gown.

The spot where John had his claim was pretty remote. It took a good hour and a half to carry the dredge in and even longer to trudge out at the end of a long workday. In spite of the gold I brought home each time, Vicky feared our gold prospecting was an elaborate ruse to cover up an affair she imagined I was having with some mystery woman. Consequently, she would frequently make the trek there and back with us and sit in brooding vigil all day while we worked.

"You'd think she'd be happy," said John one afternoon, when we were out of Vicky's earshot. "She must know you're not doing the dirty on her by now. And we're making a half-ounce minimum a day. Why doesn't she stay back in Medford and spend your earnings? It's pretty clear she hates it out here."

I shrugged. "The constant complaining and pouting give it away, huh? Actually, she's already spending my earnings. Who do you think pays her college fees?"

John whistled. "More fool you. Don't you know you've got yourself a succubus there? Sure, she's a looker, but there's more to life than that. She doesn't make you happy any more than you do her. You should get the hell away before she sucks all your life out as well as your money."

Of course, I never listened to John. In spite of our obvious incompatibility, I'd convinced myself I was in love with Vicky. We were a couple. I couldn't leave her. Besides, on more than one occasion she'd threatened to commit suicide if I ever did, and after what had happened with Gwen, she knew I wouldn't take that risk.

Apart from tinkering with the gold dredge, John had one other thing he was passionate about: permaculture. He had some land down by the Oregon coast and was always telling me how he planned to turn the gold money into self-sustaining orchards, vegetable gardens, livestock, and an 'anything else a man might need to survive' homestead.

"Ain't you tired of sucking on the teat of Babylon, Jason?" he asked one afternoon. "Isn't that why you prefer to hunt gold all day instead of getting a proper job? You should come down some time and see what I'm doing. This is real and it's going to change the world. Well, at least it's going to keep me and mine happy and well fed while the world goes to shit in a shoehorn. Why eat crap dripping in chemical pesticides and fertilizers when you can grow your own healthy food without that crap? It's not difficult, you know. Don't make any sense living the way we do."

"Does ice cream have chemicals in it?" I asked. "If it does, then I'm all for them."

"Stop joking, man. You know ice cream's full of that shit. But, you know what? It didn't use to be. And it doesn't have to be neither. Nothing does. Research it. Check out Bill Mollison. Permaculture will change the way we produce our food. It's got to. We can't keep destroying our soil like we are. Mono-crops don't work long term. GMOs are bad news, no matter what the marketing propaganda tells us. Remember that. They won't feed everyone like they say they will. Instead, they'll mess up whole eco-systems. So, what's the point of them? Profit, is all. Now permaculture is all about imitating nature. It's the only real answer to the food crises we're heading toward. Less work in it too. Once everything's set up."

I had a vague idea of what he was talking about from reading some books on the subject while in college. But as I wasn't interested in becoming a farmer, the concepts never grabbed me the way they did

John. However, the combination of his enthusiasm and the thought of how much Vicky would hate it was enough to convince me to go help him get his land set up.

Once we got started, I found the time I spent at John's land became an oasis of contentment and purpose in an otherwise grim and tumultuous time in my life. The gloominess I'd carried since childhood had deepened in my early twenties, taking root to become an integral part of my character. Playing house with Vicky hadn't helped. But I was afraid of being alone, afraid of judgment, afraid of leaving and being the bad guy.

When I was out at John's land, none of that seemed to matter. While digging swales and growing tomatoes, time stood still. I could ignore the troubles on my mind and just be.

John's land became so much of a sanctuary for me, I began going there every weekend, most of the time by myself. I built a lean-to hut at the back of the greenhouse, which became my home-away-from-home, a place to go when things got too stressful and heated with Vicky in Medford. This meant I lived in that hut almost fulltime for over a year.

I watched the spring downpours wash away our hard-dug swales, then dug them again, this time at the right levels and angles so they'd work. I planted trees and crops, and made compost, using worm castings to speed up and enhance the process. I lived and learned the practical techniques of permaculture inside out. As a result, the place bloomed. Even John was astounded at what we'd achieved in so little a time.

When Vicky graduated, the two of us moved to Alaska so I could work with my dad and save money for us to get married. Then, one afternoon while I was out working, Vicky disappeared from the apartment with all our belongings. I received an email a few days later explaining that her family had come up from California with a moving-truck. I'd figured as much, but it was nice of her to confirm it.

With the exception of the rice and beans we were supposed to have had for dinner that night, I couldn't have cared less about her leaving, or about the things she'd taken. The only emotion I'd felt the night she left was a kind of hollowness, an absence of sentiment, which could just as easily have been put down to not having anything to eat.

The day after Vicky left, I called John down in Oregon to see if my hut was still there at the farm. He told me he'd sold the place after getting a job offer in Chicago. The pay was great, he'd said, and the company had set him up with an apartment downtown, only a few blocks away from a Whole Foods store.

Disillusioned with the people in my life, I did what a lot of men do, I threw myself into work. Traveling around the state, I worked and saved money, with no idea what for.

Even though I had no garden of my own, I continued to keep the composting worms, and would give away their castings to anyone who showed an interest. This odd hobby earned me the nickname *King of the Worms* from work colleagues. Appropriate, I always thought, for someone who considered himself a dead man walking.

*

The fire in the cave died down and I drifted off to sleep, still mentally churning over my past. I dreamed of worms pushing slowly through the cold, hard ground of winter. They found my body, warm and soft, and devoured it, transforming it into castings; the raw material of new life.

Now a disembodied and dispassionate observer, I watched as my ice-white bones were dug up and labeled by a bored student, who, on closer inspection, turned out to be Vicky in a white lab coat.

Through a window, I saw a stone temple I recognized from a photograph in a classroom somewhere. The early morning sun bathed the ancient shrine in copper light. I noticed the jungle creeping in, rising in a profusion of emerald hues, until the colossal pyramid was lost in a sea of monstrous leaves.

I jerked awake. Silence and darkness enveloped me, the leaves still shrouding the dream-temple. Disoriented and groggy, I saw a feeble orange glow at my feet; the embers of the fire. I was standing. The cave stank of rank sweat, tangy piss; the smell of fear. I was nauseous, disgusted.

Someone was curled up by the fire, between the glowing ashes and Bucky. Bucky whimpered and raised his head, watching me from across the slumbering form. The ash-dusted, dying coals flickered and blazed into flame. Closer now, at eye level, I watched the tangerine light dance over the charred gray wood. I was back lying on the ground. I sat up and rubbed my eyes. The smell was gone. I was wedged between the dog and the fire. Bucky continued to whine until the flame died away. Then he sighed and fell silent, his head slumping back between his paws.

I lay awake in the dark until the alarm on my watch beeped to tell me it was morning and time to get up. Bucky ran outside while I packed up our camp. He came back dry. The storm had passed. I left the cave and stepped out into the sun, my head full of plans.

Jason Pednault   Matheu DeSilva

The Jaguar and the Monkey

When I stepped off the plane in Belize it was like walking into a furnace. It was winter back in Alaska, and the thermometer nailed to the wall of my dad's garage read minus twenty-five when we left. In Belize it was ninety-five and humid as hell. My clothes were stuck to my back and thighs by the time I'd walked down the steps from the plane to the tarmac. I couldn't have been happier. Tropical was what I wanted. It's what I'd been dreaming of ever since leaving the cave in Chugach.

I spent a month researching all the countries in Central and South America before deciding on Belize. My criteria included caves, ancient ruins, turquoise seas and jungles; all of which Belize had in spades. And unlike the rest of Latin America, the official language of Belize is English, making it the perfect place for a monolingual gringo making his first foray into the tropics.

"Sah! Ya whaa taxi?" asked a middle-aged white man with an ivory, Einstein-afro, and skin like dyed red leather as I left the terminal building.

His teeth were horsey and yellow as scrambled eggs, and his eyes were watery-gray and bloodshot. When he smiled, however, he did so with both eyes and teeth, so on instinct I decided to trust him. I nodded yes, even though the only word I'd understood was 'taxi'. I followed him to a rusty old blue Volvo and asked him to take me to the bus terminal. I spoke slowly and loudly, to be sure he understood.

He roared with laughter. "English is wha me speak, bas! Ya sound like eediat talking dat way!"

Two long bus rides and another, bumpier, taxi ride later, I arrived at a hotel at the edge of the jungle. Exhausted, I hit the bar for a couple of beers and then collapsed into bed, sleeping soundly until morning.

After a breakfast of fresh fruits, toast and eggs, I joined a group of fellow Americans and went to the nearby jaguar reserve. We hiked the trails, slid down slick and mossy limestone flumes, splashed around in the waterholes and listened to Gibraltar the guide describing the local fauna and flora. Including the habits of the jaguars that allegedly thrived within the reserve.

The lush plant life, sparkling waters, colorful birds and giant butterflies blew me away. The energizing effect I felt out in nature was ramped up in the jungle. It fizzed in my veins and made me grin like an 'eediat'.

I chose to begin my trip at the reserve because I was desperate to see a jaguar in the wild. Even though the odds were slim, I knew I wouldn't be satisfied unless I did. They were definitely around. We'd seen footprints and scat. But, by the time we had to go back to the hotel, we'd still not caught a glimpse of one.

"Y'all keep mekin big noise an so ya nah see big kitty-cat," explained Gibraltar when I asked why we hadn't made a sighting.

He had a point. Our group was loud. Even when we were supposed to be hiking the trails in silence, the constant stomp, rustle and chatter was enough to put a troop of monkeys to shame. Still, none of my fellow visitors seemed to mind that we'd missed out on the reserve's star attraction. They were all too busy talking about where they were going next to worry about where they were today.

People generally only came to the reserve for a day or two. To romp around the jungle and have fun with their fellow tourists. If they caught sight of a jaguar in its natural habitat, well, that'd be the icing on the cake, but nobody expected it. I'd booked into the hotel for a week. It was all about the jaguars for me, or at least it was in the beginning.

On day two I was already bored of the flumes and official tour stuff, so I pulled Gibraltar aside when we reached the second swimming hole to ask if we could spend more time out on the trails.

He nodded to the mass of splashing white and lobster-pink bodies. "Dem 'appy, Jason. An me lak see dem 'appy. Whaa you wan mek dem go huffin' an puffin' around in dem big lardy body instead a rollicking in da whatah lak hippo in heaven?"

"I don't know. Maybe so we'd have a chance of spotting a jaguar. This is a jaguar reserve, right?"

"True dat it is. An mebeh we see im an all im spots when we leave in a lickle while. You hear me tell dat jaguar lak tah roam in de evenin' o' early marn?"

"Yeah, I heard. But I also heard you say they get put off by noise." I glanced towards the 'hippos in heaven' and grimaced.

He inhaled his laugh, making a low, raspy sound like the bass notes of an accordion. "True dat, too! And dem get noisier day by day. Tell ya whaa, me nah ya daddy an me nah gwan mek fuss if ya whaa tek stroll bah yasel. Jus stay close an nah get loss o kill, an mek sure ta give me a nice tip fa de tip am given ya."

He smiled and waited patiently for me to decode his patois. I nodded and we shook hands, then I pulled on my daypack and walked back along the trail we'd came in on.

"Jason," Gibraltar called after me. "One hour den ya come back, ya hear? An don let a big kitty-cat crack open ya skull an munch up ya skinny lickle body."

With Gibraltar's warning and wheezing laughter still resonating in my ears, I pushed on along the track, heart hammering and adrenaline surging. Once out of earshot of all human noise, I stopped to catch my breath and feel the jungle. I was hyper-alert to every sound and movement; I heard birds squawking in the distance, trees creaking, leaves flutter. Something rustled in the undergrowth away to my left and my body tensed, ready to run or fight. I was scared out of my wits, but in a good way - I felt alive.

I moved on, and the initial adrenaline rush drained away. I was surprised at how quickly I felt at ease in what, for me, was an intimidating and alien environment. Sure, every cracking twig or movement in the foliage brought me zinging back to full alertness, but it wasn't the nerve-jangling, panicky feeling I'd first experienced. It was more of an appropriate vigilance, a base communion with the jungle, a re-awakening of something primal and necessary.

I spent the rest of the day circling back and forth between the tour group and my lone wanderings. I saw many things I'd have missed if I'd stayed with the stampeding herd; a toucan with a scarlet-tipped neon-green beak and yellow and black plumage, a vivid green parrot-snake, a giant grasshopper, striped yellow and black like a wasp. But the one thing I didn't see was a jaguar.

I didn't mind so much by now. I was too excited about the experience of connecting with the primeval jungle. Besides, I figured time was on my side; I wasn't due to leave for the coast for five more days.

My third day in the reserve played out much the same as the second. Still no jaguar sighting, but again it didn't seem to matter anymore. I was opening up to the jungle, and the jungle was opening up to me.

On the fourth day it happened.

I'd eaten lunch with everyone else and then slinked away to explore. It was a little after midday and I could still hear the whoops and squeals coming from the swimming hole as I wandered along the trail, my belly full and my mind relaxed to the point of grogginess.

There was a crackle in the bushes ahead of me, and I was sure I was about to get my second sighting of a peccary – the hairy, pig-like critters who roam through North, Central and South America. I paused and waited, wondering if this would be a lone animal, like the one I'd seen the previous day, or a small herd. According to Gibraltar, it was the wrong time of day for a jaguar sighting, so it was the last thing I expected to see.

A shockwave of electricity zapped through my body the moment I saw the stocky feline form emerge onto the path. I wasn't the only one to feel the primal jolt – the black jaguar shuddered as well, before turning its level, leaf-green gaze upon me. Crouching, it stalked toward me, head low, eyes keen and fixed on mine.

Not knowing what else to do, I froze. It occurred to me in a distant, abstract way that it might be about to pounce, and that I would be dead meat if it did. It was at least two-hundred pounds and obscenely muscular. Sleek and spellbinding, its movements were contained and fluid.

The fight or flight part of my brain had short-circuited; I wouldn't have been able to move even if I wanted to. Which I didn't. This wasn't like the bear encounters I'd had. Adrenaline dripped instead of rushing, leaving me watchful, present, but strangely unafraid.

This may have been due to the fact we were in a jaguar reserve. These animals were used to seeing, and more importantly, not eating, humans. But at the time it felt like something more; a kind of cross-species telepathy as old as time, and all but lost in the rush of the modern world. The message the panther was sending me was clear: *I could kill you today. But I won't.*

The massive cat stepped into a patch of sunlight about ten feet from me and stopped. I could see the rosette markings, pitch-dark patches beneath the burnished black of the jaguar's coat. Its eyes never left mine for an instant. It raised its nose to sniff my scent. I wanted to reach out and touch its fur. But we both kept a respectful distance.

I nodded and it appeared to nod back. The connection I'd felt with the jungle crystalized in this moment of intimacy with its apex predator. I was about to smile but checked myself. I didn't want the gesture to be misconstrued as a show of my pathetic fangs. Instead, I nodded again. This time the jaguar didn't move, it just held my gaze.

I've heard people say eyes are the windows to the soul, and I believe it. I stared into the jaguar's eyes, hoping to find some insight, some tangible, life-affirming truth I could take away with me. All I saw was mystery. As deep and green and remote as the jungle itself.

I blinked and the jaguar disappeared. It was there and then it was gone. Like a magic trick. My eyes darted left and right, making a frantic search of the vegetation around where it had stood. There wasn't even the slightest ripple of movement, and I wondered if I might be going a little mad.

The chances of seeing a wild jaguar, even in the reserve, were low, and to spot a black one lower still. I suspected I might have experienced some kind of hallucination, born of desire, an overactive imagination and the strange yellow fruit that had been served with lunch. Then I saw the pawprints in the mud.

I smiled and knelt in front of them, placing my hands inside the closest set - the jaguar's front paws. They were wide and deep. It had been one big cat, that's for sure. From this low angle, I could see where the jaguar had gone. A tunnel was beaten through the vegetation to my left, and I saw more of the fresh pawprints leading through it. It hadn't been a ghost, after all. It had only moved like one. I toyed with the idea of tracking my new friend through the foliage but decided not to push my luck.

Instead, I returned to the swimming hole where Gibraltar was watching over the tourists. He took one look at my face and broke into a huge grin. Unable to speak as yet, I grinned back and gave him the thumbs up. Mission accomplished, and then some. The buzz of meeting the jaguar moved me like nothing before. I felt exhilarated, shifted, realigned.

*

After stopping for a night in Belize City, I set off for Ambergris Caye, Belize's largest reef island, a day late. Which was perfect timing as I ended up encountering a beautiful English woman called Anna on the ferry across.

I noticed her the moment she boarded. She had a huge backpack like mine and was dressed in a long-sleeved shirt and cargo pants. Her pale skin, auburn hair and prim body language was such a contrast to the sea of bawdy blonde-streaks and sunburned flesh surrounding her.

When the ferry set off, she took out a book and began reading. I watched her until she glanced up and caught me staring. I must have looked startled; her eyes were the same shape and shade of vivid green as the panther's. She gave me a demure smile and returned to her book. I flushed bright red, and spent the rest of the crossing studiously looking out at the sea.

When we docked, Anna and I ended up walking toward the line of taxis side by side. With our matching backpacks, we looked like a couple of tourists traveling together. The taxi driver at the head of the line certainly thought so. He took her bag and then asked me where we were going. Anna flashed me a grin, then tapped him on the shoulder and told him the name of her hotel.

"That's where I'm going, too," I said. "Do you want to share the taxi?"

She raised her eyebrows and the taxi driver glanced between the two of us. "Ya nah together?" he asked.

"We are now," she said. Then to me, "Hop in."

At the hotel, we were given adjacent rooms, and we took the elevator up together, laughing and flirting all the way. By the evening, we gave up any pretense of resistance and let fate have its way. A week later I was accompanying her to the family wedding she'd come out for at Captain Drake's Resort as her 'plus one'.

*

The hotel where Anna and I were staying was pretty fancy, but Captain Drake's was at another level. The driveway alone was at least a mile long, flanked by a horse ranch on one side, and the obligatory eighteen-hole golf course on the other. The hotel itself resembled a governor's residence from colonial days, with incredible views of the ocean and cropped, shady lawns leading down to a perfectly combed beach. A marble edged infinity pool was the only concession to the modern age, and even that had an air of timeless class about it.

We'd hired a car and drove in early on the day of the wedding to beat the hot mid-morning sun, and 'to soak in the atmosphere of the occasion'. Which meant securing a prime position at the bar, from which we marinated ourselves in the free alcohol and observed the procession of fancy folk in fancy clothes and even fancier hats fill up the main lawn.

"Why don't you have one of those big English wedding hats?" I asked, sipping my third rum and coke, the sun still not quite at its zenith.

"Because I'm one of the black sheep, darling. Hadn't you realized?" I grinned woozily, which she took as her cue to continue. "English weddings are all about hats and booze, you see. White sheep wear the hats. Black sheep drink the booze." She raised her glass. "Here's to the black sheep! And to good old Uncle Jim, the big woolly ram who's paying our tab!"

"To Uncle Jim," we chorused, and downed our drinks.

The barman had fresh ones lined up before our empty glasses hit the counter. "I hope Uncle Jim has big pockets," I said, while stuffing a few more dollars in the tip jar and taking in the now crowded bar. "We're not the only ones hitting it hard."

Anna leaned in. "Of course he does. He's paying for the whole thing, silly. Not just the bar. Claire's his daughter." She glanced around then dropped her voice to a whisper. "He has very deep pockets, filled with deep-sea treasure. Uncle Jim is a bigger pirate than the chap this place is named after."

"I thought it was named after a bottle of rum. And who the heck is Claire?"

Anna laughed. "It may well have been, I hadn't thought of that. Although I'm quite sure the rum is named after the pirate. So, either way, this hotel bears the name of the gallant scoundrel to whom I'm referring. Oh, and Claire's my cousin. Otherwise known as the bride. It's her big day we're celebrating."

"Right, I think you told me that."

Anna laughed again, knocked back her gin and tonic, and waved the empty glass at the barman. Man, that girl could drink. I was already lost somewhere on the road between merry and drunk, and Anna was matching my drinks two-to-one with little to no obvious effect.

It occurred to me one of us would have to drive later, so for the late-afternoon ceremony and dinner I switched to water. Anna, like most of the other Brits present, kept chugging the booze. Gin, wine, more gin, some shots; she and her people defied the laws of physics. Not only were they still able to stand after drinking so much, they could dance and hold coherent conversations, too.

I'm not one for dancing, so after a couple of songs I sneaked away to watch Anna and the rest from my barstool. A big guy with tan skin and thick, curly gray hair plonked himself down in Anna's seat beside me. I didn't recognize him at first. His suit jacket, cravat and top hat were gone, leaving his loose-fitting white shirt to billow like a sail in the breeze.

"Scotch?" he asked, proffering a bottle of amber liquid. His hand was the same size and shade as a baseball catcher's mitt.

"Sure, thanks." I didn't really want to drink any more, but I also didn't want to refuse the host.

He poured our whiskies, then leaned back against the bar and took a deep sip. I was thankful he didn't slam his drinks like his niece. As it was, he'd decanted triple measures into each of our glasses.

"Jim Fletcher," he said, and we shook hands. Before I could tell him my name, he added, "So, how are you enjoying Belize, Jason?"

I was flattered. Not only had Anna told him my name, but he'd also remembered it. "I love it. It sure beats Alaska at this time of year."

He grinned. "Yeah, it's a bit nippy up there for my taste, too. You did well to escape."

"Have you been there? To Alaska?"

"Of course. Anywhere there's gold, I've been there. It's beautiful land. Stunning, truth be told. But cold enough to freeze the balls off a brass monkey anytime but a few days in mid-July from what I recall."

"You went there for gold?" I felt the old, familiar thrill at the word, and a sudden unexpected kinship with this wealthy Englishman.

"That's right. Didn't Anna tell you? Gold's how I made my fortune, boy."

Kinship or not, I didn't like being called 'boy', so I figured I'd give him a dig back. "Actually, she didn't mention any details, just said you were a pirate like Captain Drake."

He chuckled and raised an eyebrow. "Did she? Well, she always was very astute. I hear you've done some prospecting yourself?"

I nodded. "A little bit. But not enough to make my fortune."

"*Yet*. Don't forget the 'yet', Jason. Not enough to have made my fortune, *yet*. It makes all the difference. You're not dead. You've still warm blood in your veins, don't you? As long as you have, don't you dare talk like a quitter. And that's an order, lad!"

I laughed. I could see why Anna was so fond of her Uncle Jim. "Yes, sir, Mr. Fletcher."

"It's Captain Fletcher, actually, and appropriately. Mind you, the amount of ships I've been through it should be Admiral Fletcher by now." He caught my quizzical look and added, "One doesn't scavenge for sunken treasure without occasionally falling foul of the reefs and rocks that sank the bastards to begin with. Not if one is doing it properly, of course. Going where the big prizes are to be found. With the 'yet' always in mind."

"Are you trying to scare him or inspire him, Uncle?" Anna had appeared at my side. She was leaning up against me, squeezing my thigh. "One can never tell with you."

Jim Fletcher beamed at her. "That remains to be seen, my dear. Both, with any luck." He stood up and clapped his enormous, fleshy hand upon my bony shoulder. "Good speaking with you, young man. Take care of my niece. And remember... yet, my boy, *yet!*" He wandered away, the bottle of scotch in one hand, two clean empty glasses in the other.

Anna rolled her eyes. "Sorry about that, darling. He's so bloody drunk." She sniffed at the tumbler in my hand. "Mm, good stuff though. Balvenie by the smell of it. Probably thirty years old, if I know my uncle. Do you mind?" I smiled and shook my head. She took the glass and downed what was left. Then she drank what was left of her uncle's, too. "Mm, so, one more dance then back to the hotel?"

"Sure. As long as you let me drive."

Anna laughed and pulled me to my feet. "Let's see if your dance moves have improved first. Can't have a man with no rhythm driving me home."

*

Anna's family, and Jim Fletcher in particular, were well connected throughout Belize. Every major business seemed to be either owned or part-owned by the Fletchers, or by friends of theirs. They had a history that went back years. At certain levels of society, I could see colonialism in Belize had never really ended.

Anna had been coming to Belize since she was a small child. While at university, she'd spent her summers leading caving tours throughout the jungles south of Belize City. Then, upon graduating, she'd been hired by Uncle Jim to 'learn the ropes' of international commerce with Fletcher Holdings.

"I owe so much to him," she said one afternoon by the pool. "But I could never have stayed with his company. No way." She shook her head and looked out at the sea. "Although at times like this I wish I still lived here and didn't have to go back to dreary old London."

"Then don't," I said. "I'm sure he'd give you your old job back if you asked."

"Weren't you listening? I couldn't *really* come back and live here. That's just something people say, silly. Besides, I love my job. Fletcher Holdings was never my cup of tea."

"Why not?"

She shot me a look equal to any the panther could have mustered.

"Why so nosy all of a sudden? Can't we just enjoy the moment? If you must know Fetcher Holdings, and Uncle Jim for that matter, are too high profile, too *crass*, for my taste. He's all about the high seas, the bluff and bluster of it all. Whereas I prefer to be more low-key, subtle and mysterious." Her expression changed minutely, threat becoming flirtation. "You must have noticed how much I enjoy hiding in the dark, Jason."

I blushed like a schoolboy. "Yeah, I noticed. Speaking of which, could you show me some of the caves where you used to go? I've felt really drawn to caves lately. Probably because the open sea isn't my thing, either."

She giggled. The day before, I'd taken out a kayak and got stuck beyond the bay when the tide turned. It'd taken me hours to get back in, and I'm sure I'd only made it then by means of sheer and prolonged panic. Anna, of course, had found the whole thing hilarious.

When she stopped laughing, she said, "Okay, sure. If you promise not to fall in love with Miss Crystal Tits."

"Who?"

Anna gave me one of her sweet little smiles, closed her eyes and lay back on her lounger. "You'll see. She's a *very old* friend of mine."

*

The next day we relocated to a jungle lodge not far from the cave sites. The accommodation was luxurious, yet simple, with screen walls and covered walkways that gave the impression of being outside at all times. It was a far cry from the noisy, air-conditioned box I'd stayed in when visiting the jaguar reserve, and, with Anna's family discount, it cost a quarter of the price.

The clamor of the jungle coming alive at nighttime caught me off guard on the first evening. Anna was tired and already sleeping by the time night fell. She would have been used to it anyway, of course, but

I wasn't. I lay beneath the slowly revolving blades of the polished mahogany fan and listened to the cacophony outside, trying to decipher each and every sound with nothing but my imagination as a point of reference.

I slept fitfully, but felt rested enough by morning. The jungle was already having its energizing effect on me. Anna was in her element, too. When she returned from the kit room with the ropes, hardhats and harnesses we needed for the day ahead, she was giddy as a schoolgirl.

"Wow, and to think all I brought was a sun hat and machete."

"Yes, quite! The daft Indiana Jones hat you can leave behind," she said, tossing me one of the hard hats. "But every brave knight needs his Excalibur. You never know what you'll find lurking in the underworld."

We drove a borrowed jeep out and then hiked to a high bluff with majestic views of the jungle canopy below. Mist drifted between the tall, bushy treetops, giving the impression of squat dark-green islands in an undulating, ethereal sea.

"Are you ready?" Anna asked. She was kneeling beside me, sorting through the ropes and harnesses.

I looked around, but couldn't see the cave entrance. "What for?"

"The descent into Xibalba." She pointed over the edge of the cliff. "You didn't think your knightly quest would be easy, did you?"

"You serious? How tall are those trees down there?"

"About a hundred feet or so. Before you ask, we're going to rappel three-hundred feet down the cliff-face. Sixty feet from the bottom, the cliff's going to disappear. That's the cave entrance. There are other ways in, but none as much fun as this. Are you game?"

What could I say? *I've never done this before and I'm terrified of falling. I have little faith in third world ropes. I'm not sure I trust you know what you're doing in spite of the bright, 'this is so easy, so don't you dare disappoint me' look on your face.* In the end, I settled on, "Right on, let's do it. All part of the adventure, right?"

If I didn't regret agreeing to do it right away, I did by the time Anna had secured the ropes and we were ready to go. A steady drizzle had begun to fall while she'd worked. Now it was raining hard. She gave me a brief instructional and then stood back, nodding to the cliff edge.

"Okay, Jason. Scoot over to the edge and lower yourself back like I showed you. Once you get a little way down I'll come down and join you."

I was so petrified I could feel my legs shaking as I leaned back, inch by agonizing inch. I really didn't trust the rope until I realized my body was slanting too far back for it not to be supporting my weight. I was so relieved, I thought I'd risk a joke to save some of my male pride. "You know, for a moment there, I thought this was how you got rid of your boyfriends when you got sick of them."

Anna gave me a wicked smile and pulled the machete from my daypack, which I'd left on the ground at the top of the cliff. "It's not too late for that, you know. If you don't stop pussy-footing around, I might just cut all ties between us right here and now. Now get yourself horizontal and start pushing off. We don't have all day."

Somehow, I managed it. With heavy raindrops stinging my eyes, I kicked away from the cliff-face and let the rope slide through my hands. It was slippery and hard to grip, but once I'd calmed down, I soon got into a rhythm.

The rain continued beating down, lashing my face, obscuring my vision and making the rope slick in my palms. By now though, I didn't care. It all added to the excitement. It was me and the rope, alone against the elements.

Then a shadow materialized through the raindrops above me. It was large and moving fast. I flinched away, frightened I was going to get hit. But the dark form sailed past a few feet to my right. I caught a glimpse of rope, yellow and red, flicking out in a flurry of droplets, a dark ponytail under a green hardhat, my backpack, long slender legs in khaki shorts and hiking boots. As quickly as she arrived, Anna disappeared into the mist below.

I tried to pick up the pace, but found it more difficult to grip the rain-soaked rope, so I eased up again. Even so, it wasn't long before I reached the treetops. Here, I paused to catch my breath. I was two-thirds of the way down. Nearly there.

I felt less exposed as I dropped into the shade of the trees. My world grew reassuringly darker, less expansive. The rain wasn't so bad now either. I felt good. I was going to make it. Springing off the cliff-face, I swung out, down and back in again. Then again, and again. This last time, however, there was no cliff to hit on the return. My momentum took me in under the overhang, and the rope twanged and shuddered along the uneven ledge above. My heart skipped. I'd reached the cave.

I dangled for a moment, and then lowered myself like a spider, all the while taking in the scene around me. Anna was sitting at the edge of a wide clear stream at the cave's entrance. The stream sparkled and rippled, pale green and gold. The foliage at the river's edge on the jungle side was verdant, full of color and life. It died off, however, when it met the smooth rock at the entrance to the cave.

I splashed down into the knee-high water and unhooked my harness. Anna beamed at me from the bank. "You took your bloody time! Beautiful, huh?"

"It is," I agreed, wading over to her.

"We'll leave the ropes for the tourists Giles is bringing out here this afternoon and go back a different way. The harnesses we'll need though. Here's your bag." She threw my backpack at me, and I caught it just before it hit the water.

"Careful, my camera's in there," I complained.

"You should have thought of that before leaving it for me to carry down. Not very gentlemanly, Jason, I must say."

"Sorry about that. In all the excitement I must have forgotten it."

"I'm sure. Right, are you ready to enter Xibalba?" She held her hand out toward the cave. "It's what the locals call it. It means the underworld. A place of dread, and also hope, for the Mayan people."

I tried to look suitably impressed by Anna's spiel, while opening my pack, stuffing in our harnesses, and checking the plastic bag I'd wrapped around my camera was still secure. I hoisted the pack onto my back and held out my hand for Anna. She took it and dropped into the river beside me, kissed my lips, and then led me into the cave.

We walked in the river, which was wide and shallow to begin with. Small fish nipped my legs, and bats wheeled and fluttered overhead. The twilight created by the large entrance soon ended, and we passed into absolute blackness. We waited as long as possible to turn on our headlamps, not wanting to disturb the timeless tranquility of the cave.

We waded in silence, my headlamp illuminating Anna's back, hers the rippling river ahead. The constant drip from the stalactites beat a staccato tattoo across the water and rocks.

"I never asked," Anna whispered after some time. "What is it you love so much about caves? Do they remind you of your childhood?"

"No, why?"

"Didn't you say you grew up in Cave-something, Arkansas? Surely, they had caves there? That's usually how place names work, isn't it?"

I thought about it, trying to recall something, anything, from that part of my childhood. Nothing came. "I guess they must have. To be honest, I don't really remember much about being in Arkansas. Just one time when my dad came to visit me there on a big motorcycle. Most of my memories from then are from Wyoming." I shrugged, and a tension I hadn't noticed before eased between my shoulder blades. "I guess Cave-something in Arkansas, must have been as boring as it sounds."

Anna chuckled. "You know, for a Brit, it actually sounds rather exotic. Okay, the cave gets pretty narrow up ahead now, and the river's going to be too deep to walk. You up for a little swim?"

"Please tell me you're joking. You know I'm not much of a swimmer. Isn't there a ledge or something I can climb along? Climbing I can do."

"Afraid not. Mother Nature doesn't make things easy. But the prize *is* worth the trial." After a moment, she added, "Would you like me to help you?"

"Yes please, if you can. Maybe just help me keep the backpack up out of the water."

"I was joking. But okay, I'll rescue your precious backpack, again!" Anna backed up until she was beside me. "Right, just swim until I say it's okay for you to put your feet down again. Nice and steady. There's no rush."

I swam, with Anna behind and to my side, one arm curved under my backpack, holding its weight so most of it was above the water. The beam of my headlamp jerked around with each stroke, but Anna's stayed smooth and controlled, even though she was the one swimming one-handed. Fortunately for us both, not to mention for my expensive camera equipment, she swam with the grace and strength of an Olympian.

It didn't take long for me to get tired though. My arms ached, and my thighs and lower back began to cramp. "Do we have to swim much further?" I gasped.

Anna bobbed beside me, her arm gliding beneath the surface at infuriatingly regular intervals. "Not too far, just a few more minutes. You doing okay?"

"No, I'm not," I panted. "I'm struggling. But I'll make it."

"That's the spirit," she said, and I could hear the smile in her voice.

"Damn you Brits can be patronizing."

"I know, Jason. It's probably because we know how to pronounce the word without making it sound like the act of a benefactor. Now save your breath for swimming."

Every muscle in my body was screaming for me to stop, but I pushed through. I kept going, focusing on my breathing and counting each stroke. First, I counted to ten, but that was too painful. I reduced the number to sets of six. Six was manageable. Six strokes, then six more. Never think beyond six. Get to six, and then worry about what came next. It was hell. I was being burned alive in my own lactic acid.

After a while, I became aware of a subtle trembling along the arm snaked around my pack. It was accompanied by suppressed little snorts. Anna was laughing.

"What's so damn funny?" I asked.

"Jason, put your feet down."

I did. I could. I was speechless, torn between relief and fury.

Anna was giggling so hard she could barely speak. "I'm sorry, I did tell you about five minutes ago, but you didn't hear me. You were too busy counting and frowning."

Her laughter was infectious, and I couldn't help but join in. We were still chuckling when we reached the low, slippery ledge where Anna led us out of the river and into a large cavern.

The stalactites and stalagmites were glistening behemoths here. Ten feet tall, they hung from the ceiling and jutted from the floor like the teeth of some colossal subterranean monster. No wonder the Mayans thought these caves were entrances to a hell populated by demons and ogres.

"Take off your shoes, but leave your socks on," Anna said. "And be careful where you step." Right on cue, I stepped back on, and broke, a piece of orange and black pottery. "That's been there a thousand years, intact and undisturbed, you klutz. Try not to destroy any more priceless artifacts on our way to her chamber. Unless you want to spend the rest of your life under a Mayan curse, that is."

I shivered, and Anna raised her eyebrows.

"I've always been cursed," I explained. "With clumsiness. But I'll do my best."

The chamber was strewn with more of the orange and black ceramics, along with obsidian tools, pyrite, shells, stone figurines, ray-spines, and human bones. I was careful not to tread on anything else.

We climbed a rough ladder up to a shelf of rock, which was in fact another, smaller, chamber. Here we found Anna's 'old friend', sprawled on her back, calcite-encrusted bones thick and sparkling in the light of our headlamps.

"Wow, she's incredible. What do you think happened to her?" I asked, gazing down at the perfectly preserved skeleton.

"No one knows for sure, but she was likely some kind of human sacrifice. Maybe to Chac, the rain god. He was supposed to have lived down here with the demons. There's evidence to suggest most of the offerings this deep inside the caves correlate to a time of drought outside, so it would make sense. But there's also a theory our maiden might have been thought a witch, killed and left here unburied, so her soul wouldn't escape." I shivered again and Anna frowned. "You okay?"

I nodded and closed my eyes, trying to imagine what had taken place there a thousand years ago. There was a palpable energy inside the little grotto that gave me goose bumps. There was also a deep sense of calm, and something I couldn't quite place... acceptance, maybe, or perhaps just the weight of time measured in solitude.

In spite of the evidence – crushed vertebrae and a hatchet beside the body – I couldn't help thinking the young girl had gone to her death willingly. It was an odd thing to believe, and I had no proof to back it up, beyond some vague recollection of the nature of Mayan sacrifices from archaeology classes.

Anna and I visited several more caves during our time in the Belizean jungle. Some involved difficult climbs and swimming dark hidden rivers. Others we negotiated with ease. In all of them I experienced the same womb-like sensation of security and contentment I'd felt in the Alaskan cave. I could readily imagine why primitive societies had sought sanctuary beneath ground, no matter how foreboding the legends surrounding these dark places were.

It seemed odd to me that many people were terrified of entering caves, even without buying into superstition. It was a primitive fear that some felt and others didn't. I wasn't sure why I fell into the ranks of the fascinated instead of the fearful. Especially when, years later, my memories of *the ruin* returned.

It was only when I'd traveled further into those repressed memories that I realized *the ruin*, and what I'd taken away from it, was the reason I'd felt so at home in the caverns of Xibalba.

*

Four weeks after we'd met, and two weeks after she was supposed to have returned, Anna left for England. We said our goodbyes and parted company, never to see each other again. We both knew from the beginning that our time together was nothing more than a holiday romance; a distraction from our everyday lives. It'd been great, but it was time to move on. So, while Anna was on her way back across the Atlantic to London, I left Belize and crossed into Guatemala to see the famous ruins of Tikal.

The name Tikal means 'City of Whispers' or 'City of Tongues', depending on who's doing the translating. Built deep in the jungle, it was a major hub of the Mayan empire, boasting nearly a hundred thousand inhabitants in its heyday.

Many people would recognize the comb-topped pyramidal ruins poking through the jungle canopy as the base where the rebel forces launched their attack on the first *Death Star* in the *Star Wars* movies. Temple ruins with a sci-fi connection? There's no way an ex-archaeology major and movie-geek like me could've missed out on visiting Tikal.

I arrived at the site in the late afternoon after stowing my things at a hotel in Flores, a pretty tourist town situated on an island in the middle of a nearby lake. Tikal exceeded my expectations in every way. It was vaster, wilder, more solid and real than I'd imagined. Standing in the main square in front of the acropolis, flanked by stone pyramids over ten stories tall, I could feel the history, the lives that had played out there over a thousand years ago, before the city was abandoned. I could also feel the jungle, omnipresent and rampant, pushing at the edges of the ruins, forever ready to return them into its verdant embrace.

I wandered through the alleys and along wide causeways, passing through thick jungle to emerge again onto another clipped lawn, with another immense pyramid pushing up into the sky before me. Hieroglyphic carvings were everywhere, their meanings as mysterious and alien to me as the jaguar's eyes had been.

This remote jungle city had grown from scattered farms to an enduring complex and civilized splendor that had lasted two thousand years. Then it had all ended. The people left, and the jungle reclaimed the land, wrapping these immense constructions in moss, ivy, shrubs and trees.

I thought of the structures and institutions of our time, of our empire; the skyscrapers, bridges, banks, and churches. They appear so solid to us now, just as this ruined city and its lost civilization would have felt so permanent and immutable to its inhabitants. I was struck with a sensation of not only being surrounded by the distant, unknowable past, but also by an echo of the future. How long would it take for nature to reclaim and recycle our precious monuments?

I rounded a corner and saw a small crowd of tourists and locals gathered, staring intently at an opening low down in one of the structures. As I approached, a few of them glanced at me, but most kept their attention fixed on the dark, rectangular gap in the ruin's foundation. I asked one of the tourists what was going on.

"We're waiting for our friends to come out," he said, with a nod toward the opening. "Go check it out if you want. It's a pretty cool little cave."

I didn't need to be asked twice. I was there to explore. Cutting across the grass, I made my descent down the steep bank to the entrance. It was about five feet high and three feet wide, and dark as soot within. I reached into my jacket pocket for my headlamp, clicked it on, and ducked inside.

Before I could strap the headlamp on my head, my forehead hit something dangling from the roof. It felt like a small leather pouch, but it wasn't. It unfurled and flapped into my face, screeching, in a frenzy of crazy motion and tiny teeth. I screamed back at it, and threw my arms over my head, knocking into more of the bats as I did so. Soon, a chain reaction was set off and a swarm of the little terrors were beating their rubbery wings all around, shrieking and smashing into me.

I turned and made a desperate run for the light, the bats buffeting me as I went. The bats fluttered out after me before circling back to their dark refuge like vampires recoiling from the sun. I could hear the laughter long before the din of beating bat wings had receded.

Everyone on the bank was laughing. Some were doubled over, tears in their eyes, pointing at me and miming impersonations of my flappy-handed retreat. I laughed, too. Then took my place among the initiated to await the next poor dupe's arrival.

After I'd spent half-an-hour laughing at curious tourists being reduced to yelping wrecks by the bat-cave experience, I went to look for Temple IV, the tallest pre-Columbian structure in all the Americas. This was the one I wanted to climb.

By the time I found Temple IV, nestled in the jungle at the western edge of the site, the afternoon was almost over. The sun was setting, its last rays catching the edges of the temple-space perched atop the steep pyramid. There was an impression of molten gold; a sacred aura retained from its distant past, and a subtle demanding to be left alone to greet the night in peace. I took the hint and left, but not before booking myself on the pre-dawn guided climb to the top for the next day.

*

I awoke at three in the morning. Half an hour before my alarm was due to go off. It was raining outside. A heavy tropical downpour. I dressed and ate a breakfast of bananas and hot, black coffee, then waited on the covered veranda for the minibus to pick me up.

The minibus was late, and only half-full. Our guide and driver, Luis, told us the rain had put off most of our fellow tourists. "Sometime we arrive and cannot climb. Is very wet and dangerous when rain. We will see how it is. But if you want stay in hotel and try tomorrow, no problem. I go every day."

"You mean there's a chance we won't climb up there this morning?" I asked.

"Sure. But we will see when we arrive."

I climbed into the van feeling somewhat despondent. Nobody spoke on our way to Tikal. The others were either lost in thought like me, or dozing. The rain didn't let up all the way there, and when we pulled into the parking area, Luis wound down his window and shouted to someone in rapid-fire Spanish. A guy in a bright yellow plastic poncho appeared at his door and the two of them chatted for a few minutes.

He turned in his chair and said, "Si, my friend tells me is very wet. We try tomorrow, no?"

"Of course it's wet," I said. "We knew that before we got here. Shouldn't we at least go check it out?"

"You can go, but not me. I wait for the sun, amigo. See if rains stop then."

I cast a questioning glance around the van. Nobody looked back.

As usual, I was on my own. "Do you mean I'm allowed to go up there by myself?"

Luis shrugged. "Do what you want, amigo. But take good care. If you start you have to keep going. Coming down is dangerous. You can slip and die. Wait for the sun to dry before you come back, okay?"

"Sure, thanks for the advice, Luis. Catch you later."

I liked Luis. He and Gibraltar back at the jaguar sanctuary were cut from the same cloth. They didn't care what any crazy gringo did, as long as they got their tip. There's no way they'd get sued if something went wrong, like they would in the States, and they knew it. Latin America was still the Wild West, which suited me fine.

I pulled back the sliding door, put on my headlamp and stepped outside. The guy in the poncho clicked on his flashlight and gestured for me to follow him. He led me along a path through the jungle then up onto one of the causeways. We followed the causeway until we reached the foot of Temple IV. I slipped him a few bucks and in return he shined his light at the spot where he thought I should start my climb.

The stone wasn't as slippery as I'd expected, but it was steep. Although I didn't relish the prospect of the descent, especially after Luis's eloquent warning, the climb proved to be relatively easy. The rain eased off when I was around halfway up, and by the time I reached the top it had stopped completely. I clicked off my headlamp, sat down and waited for the sun.

The clouds broke into wisps, revealing a sky of brilliant white stars, diamonds scattered over black velvet. Then dawn drew its shroud of silver light over the heavens in preparation for Kinich Ahau, the Mayan sun god, and the stars, in their turn, faded too.

The sky to the east shimmered gold, and the tendrils of broken clouds resolved into streaks of wine and vermillion. I could see the other temples peeking through the jungle canopy. Mist drifted through leaves and around stone.

The peace was shattered by a tree shaking to my left. I hadn't noticed it before, but it was a huge specimen. Reaching to the height of the temple, its branches hung over the broad gray stones at the top of the pyramid.

A little black monkey dropped out of the tree and ran toward me, howling and barking and bearing its teeth. I recoiled, stunned by its ferocity. The thing was small, yet unbelievably loud. Its face shrunk beneath its raised shoulders, giving the impression of a vicious, wide-nosed teddy bear.

"Listen monkey," I said, feeling scared and ridiculous. "I'm not leaving until I've seen the sunrise. It's far too steep and slippery to be going down now. You're just going to have to put up with me in your territory for now."

The beast grunted and slumped to a sitting position. I turned back toward the eastern horizon, and the monkey did the same. In an uneasy truce, the two of us watched the bright orange orb rise over the jungle ruins together.

My whole trip was encapsulated in that moment. I felt so connected to the natural world; the high trees, the vines, the distant mountains, the mist and the sun, even to my new and aggressive little friend. I was more at ease than I'd ever been in my life.

Closing my eyes, I let myself drift away. I'd never meditated before, and I wasn't sure if that was what I was doing, but I felt the pull of something immense and powerful; some vast energy far beyond anything I could comprehend. I knew the energy was within me too. It was like being bigger than the sky and smaller than an ant all at the same time.

I stayed there for two hours. At some point the howler monkey slipped away, leaving me alone with my thoughts. By the time I opened my eyes and prepared to climb back down, I knew the world would never look the same to me again. My perspective had changed. My time in Belize and Guatemala had been nothing short of magical. I stood and stated my intention to the world.

"I want this life!" I hollered at the morning sun.

From a tree nearby, I could have sworn I heard the monkey laughing.

Jason Pednault   Matheu DeSilva

Fool's Gold

The first thing I did upon returning from Belize and Guatemala was to go out gold sniping. It was springtime in Alaska and the snow was melting high in the mountains. Every day, gold from the rich seams in the bedrock upstream were being deposited into the cracks and crevices where I worked. The surging currents, bitter cold and unpredictable weather, made retrieving the gold a different matter. After a week battling the forces of nature, I gave up and went back to my day job.

I knew the waterways would calm down later in the year, but they'd still be glacial. For all its deep deposits of alluvial gold, Alaska was not the place for me to make it as a prospector. The season was too short and the conditions too brutal.

Remembering the success I'd had with John on his claim down in Oregon, I resolved to try my luck in the lower forty-eight. I wasn't yet brave enough to go where I really wanted to be - further south to the jungle wilderness I'd fallen in love with.

I upgraded my prospecting kit, bought a travel-trailer, and drove down the west coast, through Canada, Washington, Oregon and California, before heading east across Nevada, Arizona and New Mexico to Texas. The prospecting was tough everywhere I went.

Nearly every inch of river, stream, dirt and rock in the United States is claimed, which means the right to any gold found is the property of the registered claim holder, and not the guy who works to get it out. To acquire a claim, one has to buy it from the government or the current holder at a price commensurate with how much gold is believed to be present at the site. Meaning anything worth anything is too expensive for a one-man-band just starting out. You need big machines, dynamite, industrial quantities of cyanide or mercury and corporate backing if you want to see a decent return on your investment.

I scouted around for open claims, but most had been stripped bare of every last flake of gold long before I arrived. Contrary to what many people in the States like to think, there really is no such thing as the land of the free anymore, especially not for a gold man.

Disillusioned, I stowed all my kit and returned to the north where my dad had set me up with a lucrative three-year flooring contract in the Yukon. I lasted seven months. When winter kicked in and the temperature dropped to around minus-eighty, I had to leave before I lost my mind to frostbite. Belize had spoiled me; I was done with the cold, and I was done with my ordinary life.

I thought about going straight back to Belize. I had contacts there, thanks to Anna, and there was plenty of gold in the Maya Mountains. According to Jim Fletcher and his cronies, Latin American countries aren't as uptight when it comes to claim laws, so I knew I'd stand a better chance of making it there than at home. Small-scale operations worked with minimum, if any, governmental interference throughout all of Central and South America.

In the end, though, I decided not to return to Belize. I wanted a new adventure. So I found a country I'd barely even heard of before, one where they only spoke languages I didn't understand: Ecuador - a place that would change my life in more ways than I could ever have imagined.

Ecuador came onto my radar largely because of one man: Fred Miller. Fred was a regular contributor to many online gold and treasure forums. He'd lived in Ecuador for three decades, and, according to his website, had made a fortune there as a treasure hunter.

Ecuador is a small country sandwiched between Colombia and Peru, on the Pacific coast of South America. Along with its neighbors, it'd been a part of the Incan Empire - one of the richest and most advanced of the pre-Columbian civilizations in the Americas.

According to Fred, most of the Inca's famous wealth in gold was mined and smelted in the remote mountains of what is now modern Ecuador. The rivers flowing down from these mountains into the lush Amazon are packed with rich alluvial gold deposits, just waiting for intrepid adventurers to harvest.

If that wasn't enough to get me excited, there were also the legends associated with this exotic land. The most famous of these being El Dorado - *The City of Gold* - which was rumored to have once stood in the dense jungle to the east of the country. Less well known, but with more evidence to support its veracity, was the legend of Atahualpa's ransom - a hoard of treasure hidden from the Spanish conquistadors by generals loyal to the last Incan Emperor, after they'd heard of his assassination at the hands of the treacherous Francisco Pizarro.

Most intriguing to me, though, was the legend of the metallic library of the Tayos Caves, where a cache of ancient tablets, crafted and inscribed in precious metals, was purported to have been discovered in a vast cave system in Ecuador's south-eastern foothills in the nineteen-forties.

If accounts from the time are to be believed, the tablets had described the course of ancient human history. A museum was established to house them in the town of Cuenca, but this incredible find disappeared off the face of the earth when the museum was burned down in the nineteen-sixties. Local legend says the contents of the library was saved; removed and returned to the remote cave system by indigenous Ecuadorians and Father Crespi, the rogue Catholic monk who'd initially discovered the treasure.

*

The more stories Fred told, the more enthusiastic I felt about Ecuador. I was ready to go. I booked my ticket, packed my prospecting gear and give notice to my boss.

A few weeks before I was due to leave, I stumbled across the website of a millionaire health-food guru who'd relocated to Ecuador. This guy was as passionate about the country as Fred, but for different reasons. He spoke about the clean water, the abundance of healthy produce, clear mountain air, and the potential for good, healthy living. Vilcabamba, the village where he lived, looked like paradise. What's more, he said it was full of expats, so not speaking Spanish or any of the other local dialects wouldn't be a problem for anyone wanting to move there.

I wrote to him, and to my surprise received an immediate reply inviting me to stay in his guesthouse. The invitation came with the caveat that I check out the river adjacent to his property for gold, and split anything I found with him fifty-fifty. He told me he'd found gold flakes in the riverbed sand but was too busy with his health-food empire to fully explore the potential income stream – I pardoned the pun - right on his doorstep.

Everything was falling into place. Fred had reams of advice and leads for me to check out, and Ken Kerr, the health-food guy, had decided to take me under his wing and hand me a stress-free opportunity for my first prospecting gig.

*

I flew in to Quito, Ecuador's capital city, on a wet, windy February night. Quito is nine thousand feet up in the Andes Mountains and a few miles south of the equator line. The altitude and unique geography make it prone to unpredictable weather; freezing mists, howling storms, hail, hot sun and blue skies, Quito makes up for the equatorial lack of seasons by giving samples of them all in one day.

When I awoke on my first morning in the city, I opened the curtains, and saw nothing. A dense white fog had swallowed Quito whole. By the time I'd finished breakfast, however, the sun had burned away the morning mist to reveal a sight unlike anything I'd ever seen.

A multitude of sun-bleached concrete buildings in faded pastel shades stretched across the valley from the city's eastern slope beneath my hotel window. Parks, tall trees, modern shopping centers and old stone buildings punctuating the sprawl through the valley's center. Dark gothic spires, creamy colonial edifices, and a hill with what looked like a statue of an angel on top dominated the southern skyline. To the north, skyscrapers skirted a wide parkland, before giving way again to the low-rise concrete hodgepodge that filled every blank space in the city's canvas.

North and south, Quito was a never-ending muddle of old and new, an agitated mix of decadent, elegant and squalid construction. East and west, however, the jumble was hemmed in. The city was only ever a mile or two across, stopped from advancing further by the colossal green flanks of the city's main feature: Mount Pichincha - an immense and active volcano, with craggy, steel-gray, snow-dusted peaks. Pichincha stood sentinel over the city, determining its western border.

I spent twenty minutes gawping at the view, experiencing an incredulous thrill each time an airplane roared into the city valley, skimming over the buildings toward the airport in the north. Some of the planes flew so low I could almost see what the people onboard were having for breakfast.

Even though I was fatigued from the travel and altitude, I put together a daypack and set off for Fred Miller's apartment for our pre-arranged meeting. I was excited to see Fred. He was an old-school adventurer with an encyclopedic knowledge of Ecuador's gold and treasure sites. What's more, he was willing to share his knowledge with me - beginning with this meeting about the sites he'd discovered down south, where I was headed after Quito.

I imagined Fred to be a less arrogant, more benevolent version of

Jim Fletcher. An ally and possible mentor. We'd been corresponding regularly, and aside from the promise of help upon arrival, he'd already recommended some excellent books to help with my preparation. One of which was about a successful Texan treasure hunter, in whose footsteps along Ecuador's coast I intended to follow, both figuratively and literally, once I'd made my first strike down in Vilcabamba, and explored the sites Fred and I were due to discuss today.

The morning heated up fast and I had to give up on my plan of walking to Fred's apartment. Without a hat, my brains would be baked by the time I got there. I hailed a taxi and gave the driver the address. He dropped me off a few minutes later outside Fred's apartment building in a upscale part of town, a few blocks from the swanky Hotel Quito. I was half an hour early, so to kill time I went across the street to an Internet café, where I could sit in air-conditioned comfort and check in with family and friends.

Once I'd selected a booth and signed in, the first message I saw in my inbox was from Fred.

Jason,

Apologies. I'm going to have to cancel our meeting this morning. Something big is happening and I have to go away for a week or two. Top Secret. I'm sure you understand. Go to Palanda, south of Vilcabamba. You'll find gold in the rivers there, and great coffee in the cafes!

Good luck,

Fred Miller - Adventurer

Call me naïve, but in all the excitement of planning my trip this was the first time I suspected Fred Miller might be a bullshit artist. Top Secret? Give me a break. Still, the books he'd recommended supported his assertions that there was a fortune to be made in Ecuador, either with or without his help. Besides, I already had Ken Kerr's river of gold to get me started.

I left Fred's fancy neighborhood behind and went straight to the bus terminal and booked a ticket on the overnight bus to Loja, the closest city to Vilcabamba. Twenty-four hours later, I was five-hundred miles away from Quito, drinking coffee in a juice bar, surrounded by North Americans earnestly discussing conspiracy theories and alien abduction.

"Raw food or free energy?" asked a man from the next table.

It took a moment to realize he was talking to me. When I did, I met his inquisitive gaze with one of bewilderment.

"Well, you don't look like the on the run from the C.I.A. type," he continued, nodding toward a wild-eyed and shoeless old hippy a few tables down, "so I figured you must be here because of Ken Kerr, or Brian the astronaut up at Monte Sueños."

Something in his tone made me unwilling to be pigeonholed in either camp, so instead I decided to own my new identity. "I'm a gold prospector and adventurer. Here to pan the local rivers."

"Really? You think there might be gold here?"

"Why not? You've got an astronaut."

He laughed. "Yes, we have all kinds of weird and wonderful here, from Breatharians to Lemurians. I'm Ravi, by the way. I came down here for my health. I had a chronic illness. The doctors in the States wrote me off as a lost cause; untreatable, they said. Long story short, diet and lifestyle change, coupled with natural therapies, worked better than the best modern medicine had to offer. Who'd have thought it, huh? Not me, that's for sure. But here I am, living proof.

"I'm healthier than ever, which is just as well because I have a wife and son who need me. I hope you find your gold, man. Let me know if you ever need help. In general, I mean, not with the gold panning. I know how hard it can be settling in to a new place."

I thanked him and introduced myself properly. We chatted for a while, mostly about his son, Ezra, of whom he was very proud. After some time, Ravi's wife Dana joined us. She was pretty and petite, and as friendly and welcoming as her husband, if somewhat less chatty. She carried a basket overflowing with colorful produce from the local market, which she offered to share in the form of a dinner invitation.

"What an excellent idea," Ravi enthused. "Dana's an amazing cook, and you, Jason, can entertain us with stories of your adventures!"

I was about to agree when Ken Kerr pulled up in a sparkling silver F350 pickup. With the engine still running, the passenger side window slid down and he leaned across to peer at the people gathered outside the juice bar. He scanned the small crowd until his eyes met mine.

"Jason?" he yelled.

I nodded.

"Let's go. Your room at the guest house is ready." The window slid up and he stared straight ahead, waiting for me to get my things together.

"Sorry guys," I mumbled to Ravi and Dana. "Dinner sounds great. Thank you. Maybe another day?" I glanced at Ken Kerr's flashy truck. "I don't know how he knew I was here. I wasn't due to arrive for a couple more days."

A thick-set blond man sitting nearby chuckled until the laugh turned into a spluttering cough. When he recovered, he spoke in a rasping rock-star English accent. "Welcome to Vilcabamba, mate. You fart in the juice bar and within an hour the whole village has heard you shit yourself."

Ravi shrugged. "I wouldn't have put it quite like that, but the sentiment is true enough. It *is* a small village after all." He shook his head then smiled. "Have fun with Ken. He's an interesting one. I'm not sure, but I think you'd prefer the astronaut. He's a little more down to earth."

"The astronaut's more down to earth – good one Ravi!" said the Keith Richards sound-alike. I collected my bags and climbed into Ken Kerr's hermetically sealed world.

Ken Kerr was nothing like his online persona, whom Thelma had described after seeing one of his videos as the hyperactive, charismatic bastard love child of a flame-haired dwarfish alchemist and a nature-loving hippy witch. I wasn't sure if she'd meant it as a compliment or an insult, but it was definitely fitting. In person, however, he was petulant and awkward. A lot more concerned with work and making money, than living close to nature and connecting with people.

When he showed me to my room in the guesthouse, he explained the rent would be three-hundred and fifty dollars a month, but not to worry, as I could pay with the gold I was going to pull from 'his' river. It turned out I had roommates, too. All of them tied in to Ken Kerr's empire one way or another. People he'd enticed down from the States and Canada to help grow his health-food business.

Kitty, a graphic designer from California, had the room next to mine. Aside from her professional skills, she was also a self-professed psychic. I took her claims with a pinch of salt at first, but when I saw how accurate her predictions were, I couldn't help but marvel at her spookiness. She was funny and I liked her a lot.

Kitty and I shared the guesthouse with two couples. Each had a young daughter. One of the guys was a genius nutritionist from Texas who looked like a hobbit. He was married to a tall, graceful elf from the western isles of Canada. Between them, they already ran their own successful business; cleansing and healing their clients through a combination of super-charged juices and profound life-coaching. I'd like to have gotten to know them better, but they left after a dispute with Ken a week after I arrived. Kitty had predicted the day and hour of their leaving on my first day there.

The other couple, a Canadian Internet marketer and his highly-strung British wife, didn't last long either. It appeared Ken struggled having any other alpha-nerds in his territory.

After I'd taken some time to settle in, I went to pan the river Ken had described as being full of gold. It came down from the nearby mountains of Podocarpus National Park and passed through the village, before winding its way out to the gated community where Ken and the more paranoid of the ex-pats lived. Out of ten pans, I managed to find one small speck of gold. What Ken had mistaken for gold was actually pyrite, otherwise known as fool's gold.

When I told him there wasn't any gold in the river, he narrowed his eyes at me. "What's all that stuff glittering in the sand then? Don't think you can hold out on me, Jason. I'll know if you're keeping it all for yourself."

I was furious, not just with his attitude, but because Ken, like Fred, was turning out to be another dead end. "Gold doesn't glitter, it shines, Ken. If you'd told me the gold you found *glittered*, it would've saved me from wasting my time and money coming all the way down here."

Kitty, who'd been working in the corner of the room, looked up at this point. "You didn't waste your time coming here. Vilcabamba is important to you, Jason. It'll help unravel your mystery."

"That's great Kitty," I snapped at her. "But there's no gold or treasure here, right?"

Kitty shrugged. "No gold, but treasure, that's a different thing. You're going to find plenty of gold in this country, Jason, but nowhere else you could call home. You'll always come back here. You'll need to eventually." I must have looked skeptical, because she smiled and added, "Not right here, you idiot. Not to Ken's place. You'll leave here today. But the village is another matter."

I couldn't help but laugh. "Was the part about leaving today psychic divination or cold-reading?"

She winked at me. "Both."

I turned to Ken and held out my hand. "No hard feelings. You weren't to know. Fool's gold has fooled plenty over the years."

Ken shook my hand but didn't say anything or look me in the eye. He left the room, and as soon as he got outside I heard him yelling at one of his local workers.

Kitty raised her eyebrows. "I won't be far behind you."

"I could've told you that. Now where the hell am I going to find all this gold you were talking about?"

She gazed out the window for a full minute, then sighed and said, "I can't tell you. I don't want to be responsible for what happens. It's going to be very bad." She paused to frown, then added, "Actually, it'll be kind of good, too."

"Well, that's a relief, or maybe it's not. Way to cover your bases. You should get a slot on a TV show, Kitty."

I packed my bags and moved to a hostel in the village. With the exception of a weeklong excursion south to Palanda, where I explored some caves and pulled a worthwhile amount of gold from one of the rivers, I didn't leave Vilcabamba for two months. I hung out at the juice bar with Kitty in the mornings, and Curly's - a real bar - in the afternoons. At least one night a week I'd go to Ravi and Dana's for dinner.

Like countless others, I'd been snagged by the Velcro-bamba effect. The village is full of people who were 'just passing through' years before, but still haven't quite gotten around to leaving. It took Henry, another contact from an online treasure forum, arriving in Ecuador from Florida, eager to meet up and go gold hunting, to get me motivated to move on again.

"Any ideas on where we should go first?" he asked over the phone from Quito. Fred Miller was apparently still on his 'top secret' mission and had consequently been unable to make his meeting with Henry either.

Speaking about Fred reminded me of the Texan treasure hunter in the book he'd recommended. "Let's go to the beach," I suggested. "If we're going after Incan gold, we might as well start with the original El Dorado."

I met Henry a few days later in Bahía de Caráquez, a pretty city in the coastal province of Manabí. From there we took a small speedboat across the bay to San Vicente, a much less pretty city, and caught a bus north. Our destination was the small village of Coaque, a nondescript cluster of clapboard huts and concrete boxes on the dirt road, south of the river bearing the same name.

According to the book I'd read, there was once an Incan city a few miles up the Coaque River. In 1531, the conquistador Francisco Pizarro had raided this city and plundered an enormous amount of gold and emeralds from its inhabitants. So much treasure was found, and taken, that the Spaniards called the place El Dorado - The Golden City. It was the nominal forerunner to the El Dorado of legend, purportedly hidden deep in the Amazon jungle.

During the raid, the Inca had reportedly abandoned their homes and taken to the forests to escape the Spaniards, never to return. Their city was then forgotten about until the 1970s, when the Texan treasure hunter discovered references to it in the British Library. He reasoned that in a city of such abundant riches, there would still be treasure-filled 'tolas' - distinctive Incan burial mounds - waiting to be discovered in the dense jungle along the river's edge.

And he was right. Not only were there gold artifacts in the tolas he found there, the whole area was awash with archaeological treasures, from crudely cut emeralds to ornate ceramics. He led two expeditions to the site and removed a number of artifacts, which he then smuggled out of the country with the help of British tourists he'd enlisted to help with the excavations.

The Ecuadorian authorities were unhappy about losing such valuable relics to a gringo treasure hunter, and on the second expedition, the Texan was lucky to have escaped the country. Still, the Ecuadorians ultimately benefited from his find. On the afternoon I'd spent in Quito before going to Vilcabamba, I'd seen gold masks and intricately carved figurines from the region in the Central Bank Museum.

I wondered how thorough the Ecuadorians had been in their searches. If the book were to be believed, the site of the old city was huge; ceramics and emeralds and other artifacts were often found washed up along the riverbanks and local beaches. The remoteness of the location and the harsh conditions of the jungle gave me hope there may yet be some bounty left for Henry and me.

We stepped off the bus into a ghost town. The only sign of life was an emaciated dog chomping on a greasy blue plastic bag. The bus kicked up a cloud of dust and diesel fumes as it pulled away, disappearing from sight before the air had cleared and leaving us alone with the dog and the silence.

"Are you sure this is the right place?" asked Henry. "The driver didn't seem to want to stop here."

"He didn't want to stop anywhere. Haven't you noticed the hurry they're all in? According to the sign back there this is it. Let's go check out the river. Who knows, we might find a nest of emeralds the size of duck eggs in the reeds."

We traipsed across the dry packed mud toward the river, bent under the weight of our backpacks. It was mid-morning, and the muggy silver clouds stretched the intensity of the blazing sun across the entire sky. The glare was blinding and the heat suffocating. When we reached the riverbank, Henry took off his pack and collapsed on top of it. We gazed down at the muddy river. Like everything on the coast except the buses, it moved slowly.

"See any emeralds?" I asked.

"Not from here," said Henry. "I don't know about you, but I'm ready to dive in to that creeping caramel and take a closer look. It's hotter than Satan's scrotum up here." He nodded into the haze upstream. "Maybe this guy knows where all the treasure is. Or, even better, the location of the nearest bar."

A round-faced man in a straw hat was approaching along the river's edge. He was leading a mule by a thin, dirty rope. He wore long pants, cinched at the waist with string and tucked into rubber boots. In his right hand he held a machete, the flat of the blade resting against his shoulder. The mule's stomach was bloated and its grimy-white coat was sprayed purple all up one back leg and along its haunch.

Flies congregated on the painted leg. Each swish of the animal's tail sent them swarming up and around only to return again to repeat the process moments later. When man and mule drew closer, I saw why the flies wouldn't leave it alone. Beneath the disinfectant purple paint, open sores oozed dark and wet and raw.

The man raised a hand in greeting. "Buenos días. ¿Están de paseo o buscan tesoro? ¿O talvez están perdidos? ¡A lo mejor las tres cosas! ¡Como todos los gringos!"

Whatever he'd said tickled him; he laughed so hard he had to dig his machete into the ground and lean on it to catch his breath and recover. Henry laughed, too.

"What'd he say?" I asked.

Henry had been in the country barely a week, but his understanding of Spanish was already far better than mine. I figured he must have learned it at school or from the Latinos in Miami where he lived.

"Something about looking for treasure and being lost, like all gringos," Henry explained, still smiling.

The man nodded at Henry's translation. "Soy Stalin," he said, pounding his chest. He grabbed Henry's hand, then mine, giving us both firm handshakes. "Mucho gusto! ¿Con que les puedo ayudar, caballeros?"

Henry's grin grew even wider. "His name's Stalin, like the Russian dictator. And he wants to know if he can help us."

"Ask him about the tolas and the old city. See if he knows where it is." I was getting excited. Maybe Stalin remembered the Texan or the government digs. He could feasibly have worked on them both. They would have used local labor.

"I'll try," said Henry. "¿ El Dorado? ¿Viejo ciudad? ¿Los Incas? ¿Tesoro?"

Stalin smiled, gold molars gleaming in the back of his mouth. "¿Necesitan guía, amigos?"

"Guía. *Guide*. He wants to know if we need a guide," said Henry.

Stalin nodded and pounded his chest again. "Si, Stalin es guide."

"We probably do," I conceded. "And I don't see anyone else applying for the position."

Henry looked over Stalin's shoulder at the mule, then back to the empty village. "Good point. Let's hire him." Then to Stalin, "Si, tu nosotros guía."

Stalin frowned in concentration. I guessed my new friend's Spanish wasn't quite as good as it sounded to me.

"Dinero?" Henry added, taking out his wallet.

He passed Stalin a ten, which, as is often the case in foreign countries, seemed to eliminate all misunderstanding. Stalin smiled, turned his mule around, and beckoned us to follow. "Vamos a la casa de Atahualpa. ¡El sabe todo, amigos!"

"He said he's taking us to Atahualpa's house," Henry said, chuckling and elbowing me in the ribs. "We're going straight to the top dog, to the last Incan Emperor!"

Henry pulled his backpack on again, and we fell in step behind our guide. In the distance, beyond the cracked bare land where we were walking, the mist parted for a moment, and I caught my first glimpse of the dense green cloud forest on the slopes of the mountain beyond.

It was past midday when we reached our destination – a bamboo shack up on stilts overlooking the river. An enormous white and black hog was rooting in the dirt beneath the house. Chickens pecked at the baked earth in the yard. A pretty young indigenous girl reclined in a hammock tied between two skinny coconut trees by the gate, her dexterous fingers pinging dried corn off the cob and into the plastic bowl in her lap. She didn't look up until Stalin hailed her.

When Stalin explained why we were there she put aside her corn, careful to cover it so the chickens wouldn't get at it. She went to the house and stood at the bottom of the rickety stairs leading up through the floorboards of the cabin. She called to whoever was inside, listened for the response, then beckoned us over. By the time we reached her, an old man had ambled down the steps to greet us. This was Atahualpa. Not the last Incan Emperor, of course, but an important man in our quest nonetheless.

Atahualpa led us to an area beneath a shade tree where five plastic chairs, all in different colors, were set around a low wooden table. Henry and I took off our backpacks and laid them against the tree, then joined the two older men. We were both tired and thirsty from the long walk and were delighted when a few minutes later the girl brought over a pitcher of homemade lemonade. There was no ice, but it was cool and sweet and just what we needed.

"I am Madeinusa," she said. "I am speaking English. My grandfather no speaking many words English. But I am translator for him. You want to know about treasure of Coaque?"

Henry and I leaned forward, nodding as one.

It turned out old Atahualpa had been with the Texan when he first discovered the ancient city. He'd also been the leader of the local workers for the second, and final, expedition. He said the Texan had treated him and the other laborers well, paying them fairly and on time, which was more than could be said for the government people who came a year later to finish what the Texan had started. Atahualpa shook his head as he told us he'd seen more gold come out of the ground over the years than he'd ever thought possible.

"Did the government archaeologists get everything?" I asked.

Madeinusa translated my question. Atahualpa laughed before giving his response.

"They think they did," said Madeinusa. "But grandfather knows different things."

Atahualpa explained how he didn't show the government people everything, because they paid him and the other workers so little. The government took a lot of gold out of the ground, but when Atahualpa judged they were satisfied, he told them there were no more sites to be explored. Apparently, they relied on local scouts like Atahualpa and his brothers to find the tolas, as well as to dig out the treasure for them.

He then waited for the Texan to come back so he could lead his friend to the hidden tolas. But he never came. I told Atahualpa the Texan had died in a plane crash some years ago. He seemed genuinely saddened by the news, but also impressed that the Texan had died while flying through the sky.

Other foreigners had come to seek the Incan treasure over the years. He assumed they, like us, were friends of the Texan (I didn't correct his assumption). Some he helped, others he didn't. It all depended on how well they treated him and his people. The ones who paid well always went home with treasure.

"We'll pay well," said Henry. "Can you take us to the tolas?"

Madeinusa explained her grandfather was too old to lead men into the mountains now, but if we wanted to go to the tolas, her uncles could take us. Stalin cleared his throat and gave her a meaningful look.

"They go with you and Señor Stalin, who is your guide," she added, smiling and nodding to each of us in turn.

That night, Henry and I were too excited to sleep. Instead, we lay in the hammocks we'd strung up in the yard, talking and planning. Neither of us could understand why the local guys hadn't taken the gold for themselves. Atahualpa had led most of the digs. He must have handled millions of dollars in artifacts, yet he lived in poverty. Quite literally a few steps up from a pigsty.

"You know how much the Texan paid these guys back in the seventies?" I asked Henry. "Forty cents a day! And God knows what kind of pittance the government paid after that. We should give them a decent wage, though. How much do you think is fair?"

"Considering inflation, at least two bucks a day," he said, laughing. "And at that rate I could keep Stalin on as my own personal butler when we leave. I already gave him a ten! Seriously though, you're right, we should look after them. Let's give them whatever they ask. Without haggling. After all, they're going to make us rich, my man!" In the end, we agreed to pay both of Atahualpa's sons thirty dollars a day, and Stalin forty, because his mangy mule was packing in our gear.

The three men led us up into the mountains early in the morning where we spent most of the day bushwhacking between various tolas.

Every one we came to had the telltale caved-in sides, evidence they'd been dug out and raided already. In spite of what Atahualpa had told us the day before, not one had been left alone.

We made camp in the evening, feeling as dejected as we'd been excited the previous night. To add to our misery, once the sun slipped beneath the horizon, swarms of hungry mosquitoes emerged from the dank forest to feast upon us. They ignored our guides and went straight for Henry and me. We were soon covered with so many of the itchy little bites that sleep became impossible. It was torture. Then it rained. Hard.

In the morning, we awoke soaked and scratching and angry. Henry was about to confront our guides about the lack of virgin tolas when Milton, the brother who spoke the best English, came to us.

"You ready to see treasure now?" he said with a wide grin. All but three of his visible teeth were made of gold, and he wore a faded blue bandana on his head, giving him the aspect of a pirate. "It's too much rain and mosquitoes here, no?" he added, with a good-humored cackle.

"Yes," I said. "We're ready."

"Yes," said Henry. "We don't mind about the rain and bugs. We'll stay here to get the treasure."

"Wrong, amigo. No need. Let's go. Body gets warm walking not talking."

We followed the three men back down the trail we'd ascended the day before, our bodies aching as if we'd spent months in the wilderness. By the time we reached the plain over the river valley, the sun had broken through the mist. We could see the ocean in the distance, glinting like a sapphire. And closer, nestled in a bend at the river's edge, Atahualpa's house.

From this vantage point I could see the palms ringing his compound, and a huge cornfield out back I'd not noticed before. There were papaya and banana fields beyond the corn, too. I smiled, thinking of those old tales where the treasure one seeks turns out to have been right at home all along.

"Where are we going?" Henry asked our guides.

I pointed to Atahualpa's house. "Down there."

The three Ecuadorians laughed and nodded. Milton slapped my back. "To the treasure, amigos!"

We sat at the same table under the same shade tree as we had two days before. This time, though, the table was filled with artifacts. No gold or emeralds, unfortunately, but a lot of exquisite ceramics - finely detailed figurines, cups, pots. All of which were over five hundred years old, and in perfect condition.

"This stuff is amazing," said Henry, turning over a cup in his hand, and then reaching for the statuette of a man crouched like a jaguar with his tongue sticking out. "How much do you have?"

"How much you need?" asked Madeinusa, flashing Henry her sweet smile.

"Why did your grandfather send us up the mountain?" I asked. "Is there anything left up there?"

"He did not send you. Señor Henry asks to go. Many foreign people like this adventure," she said with a shrug.

Atahualpa cleared his throat and leaned forward. He began speaking to me in rapid Spanish. When he finished, he kept his eyes on mine while Madeinusa translated.

"Atahualpa tells you our people are not stupid. We know many Americans and mestizos get rich from graves of *our* ancestors. He watches them. He works and learns to do what they do. It takes much time without the gringo machine that tells where to dig. But he has much time. He and his brothers take gold and sell in Pedernales. Every gram is pay for by weight. Fair price. After, they all can buy land. He keeps ceramics for gringo treasure hunters. If you pay, you have treasure. Just like he say two days ago. Grandfather is honorable man. More honest than governments and gringos."

I smiled and shook his hand. It didn't matter that he and his brothers could have made much more than the weight price for the artifacts they'd recovered. They'd still made a lot more than they would have digging on day wages for people like me. I had to admire him for that.

I only bought a few pieces, as by now I was running low on cash. Henry bought a lot. We figured our trip up the mountain had earned us the right to haggle a little on the prices, so we did. Once our business was concluded, we toasted each other's health and good fortune with a glass of raw sugarcane alcohol. I had an unsettling flashback of Everclear the moment it hit my mouth. The stuff was nasty, and powerful as rocket fuel.

Through the fog of instant drunkenness, Milton loomed into sight over Atahualpa's shoulder. The two men spoke. Then Milton addressed Henry and me. "Amigos, if you want gold, there is another place your friend finds. Many more Tolas are there. But we need machine to find the digging places. We can't wait like the old men did. You can get one, we go together."

"Sure. Let's do it," said Henry. Then to me, "I'm guessing they mean metal detectors. I can pick up a couple in Miami when I go back to sell these pieces. We could be back here in a month, Jason. Then we'll make some real money!"

I agreed to the plan and we all raised another glass to our partnership with the guys from Coaque. Shortly after, I went to find my hammock as I wanted to take a nap before the hooch knocked me unconscious. I slept through until night, when I awoke to the sight of a star-filled sky.

The stars on the equator were different. Even those familiar to me from the northern hemisphere weren't where they'd usually be. I gazed at the constellations I didn't recognize, seeing uncharted territory where old rules and limitations no longer applied. I took a long drink from my water bottle and contemplated where these new stars might lead me, until the pounding in my head got so bad I couldn't think about it anymore and I curled up and went back to sleep.

In the morning, Henry and I walked back to the highway. We caught a ride in the back of a pickup to a little beach town we'd passed on the way up from San Vicente. We stayed there a little over a week; surfing, drinking beers and making plans. It felt good, especially being so far away from all that's familiar, to hang out with a friend who not only shared my dreams, but had the courage to follow them through, too.

We had a shock on our second day in town when we discovered an antique store selling local artifacts at a fraction of the prices we'd paid in Coaque. After giving them a close examination, Henry pronounced most of the store's wares to be fakes. Which didn't make us too popular with the owner, who stood in the doorway, arms folded across his chest, scowling at us, every time we walked past thereafter.

With nothing better to do, and with my cash reserve dwindling, I decided to go to Quito with Henry when he caught his flight home. I wrote to Fred Miller, asking to meet up again as I needed something profitable to tide me over until Henry came back with the metal detectors.

Again, Fred was too busy to meet me, or so he said. But he also said he was pleased Henry and I had met up, and was impressed we'd explored Coaque together. We showed potential, by all accounts, and as such he threw me another bone: the location of a proven gold-rich river a day's ride east of Quito.

On this occasion, he gave detailed instruction of the exact location I should explore. He told me a friend of his had been extracting large quantities of gold up until a few months ago, when he met his unfortunate demise in a freak accident.

Stepping into a dead man's shoes didn't seem like the best of omens, so I wrote back asking for more details. Fred told me the man had hit a rich vein of gold on a shelf of sand surrounded by large rocks. The more he followed it, the more it produced, until he arrived at the foot of a boulder where the deposit appeared to be equal parts sand and gold dust.

In his excitement, he called Fred to boast of how many ounces he'd pulled out the seam, and how much he was going to get from beneath the boulder the next day. Fred warned him to be careful, reminding him of the Jatanyacu legend, but the man didn't listen and the next day he disappeared.

A few days later, Fred took a trip to Tena - the town where the man was lodging between forays into the jungle - to ask if anyone had seen him. Nobody had. In the end, Fred went out and searched the Jatanyacu River for the site where his friend had been prospecting. Sure enough he found the treasure hunter's body, crushed beneath an enormous boulder, pan, shovel and mercury kit still nearby. The location was remote and the boulder too large to move without heavy equipment, so he'd had to leave the guy there for the river to take him.

It was a great cautionary tale, but by now I wasn't sure if I could believe anything Fred told me. Still, I had no better options, so I decided to go check out the site for myself anyway. Before I left Quito, I wrote Fred another email to ask about the Jatanyacu legend he'd mentioned. I didn't receive his reply until a week later. By then I'd already fallen in love with the river and the jungle surrounding it.

The legend was as simple as it was sinister: *If an outsider puts a foot in the Jatanyacu, the river remembers, and when she's hungry, she will devour him.* I felt a chill run through me when I read the words, but then laughed it off. My experience of the big river, which was what Jatanyacu meant in Quichwa, the local indigenous dialect, had been serene and pleasant. Not to mention very lucrative.

At this stage, however, I'd yet to meet her when she was hungry.

The Jatanyacu

When I first arrived in Tena, I took one of rural Ecuador's ubiquitous four-wheel drive pickup taxis straight out to the Jatanyacu. To my surprise, the taxi driver spoke very good English. He'd worked for one of the big American oil companies drilling in the Amazon basin and had taken full advantage of the language lessons they'd offered to the local workers.

"The company looks after us good, man," he enthused in his faux New York accent as we bumped our way further and further into the jungle. "Lot of work for us and lots of money for the country. Tena was just mud streets before they found oil here. People forget that. And now the stupid Indians are complaining about pollution in the rivers. I mean, everyone knows you have to crack eggs to make an omelet, right?"

"Maybe they didn't want the omelet in the first place," I suggested.

"Ha! You're crazy. Everyone loves omelet. And if they don't, they should. Omelets are progress! It's cold Coca-Cola on a hot day, concrete roads and computers in schools. The real thing! These guys just don't know what it is to be civilized."

We emerged from a tunnel of green boughs, onto a ledge overlooking the river. A flurry of red and blue parrots flew up and away across the verdant canyon. Far below, the river glinted in the morning sun. "Lucky them," I said.

"Yeah, sure thing, boss." The driver shook his head. "I see the gold kit you're bringing out here. You didn't come for the sightseeing. You're here for the money. Don't pretend you're not. But be careful, my friend. Gold makes you crazy. A gringo like you came out here and dug his way under a rock a few months back. Got killed stone dead. All because he loved that yellow and green too much."

I nodded, thinking maybe Fred wasn't so full of shit after all. "Thanks for the advice. I'll try not to lose my head like he did."

The driver giggled like a little girl. He kept laughing and repeating, 'lose my head,' and 'stone dead,' over and over until we reached the end of the road. I arranged for him to come back in five days to pick me up and then went exploring.

There were a few wooden huts at the trailhead, and an old beaten up sign welcoming me to the Llanganates National Park. In one of the huts they were cooking fresh fish and yucca root to sell to tourists, though there were none there but me. I bought a plate and ate it next to a wide rock pool fed by a series of cascading waterfalls. Afterward, I crossed a rope and plank footbridge over the pool and followed the trail into the park.

I walked a long while, taking in the sights and sounds; it felt so good to be back in the jungle again. After going a fair distance, I descended along one of the dry creek beds crossing the trail, and made my way down to the river's edge.

My first thought when arriving at the river was I couldn't have picked a better spot. The beach was strewn with large flat rocks, crisscrossed with deep crevices. The river flowed slow and deep at the center of the channel, but near the edges it bubbled over the rocks, creating little rapids and pools. The giant slabs of stone attracted butterflies of every imaginable color and size. At my approach, they whirled into the air as one, surrounding me in a fluttering rainbow of delicate wings.

I made camp in the shade some distance up the bank. Stringing my hammock between two moss-covered trees, both of which teemed with orchids. Some were blooming in vivid scarlets, burgundies, golds and blues, while others were still hidden within delicate buds, their unique and exquisite splendor yet to be revealed.

My days on the Jatanyacu consisted of bathing in the river, digging out the crevices, panning out the gold, fishing, eating and sleeping. I was in paradise. And I was making bank.

Every five days I returned to Tena, to cash in my gold and spend a couple of days buying supplies and socializing with the expats and tourists, before returning to the blissful solitude of the Jatanyacu again. Everything was perfect. The longer I spent on the river, the more I believed the ominous legend wasn't meant for me. After all, I was no longer an outsider; the river and I were friends.

Then I got greedy, and everything changed.

Henry was dragging his feet coming back with the metal detectors for Coaque, but I wasn't worried; I was doing well enough in the jungle by myself. When I thought of Henry, though, the excitement of putting something big together was always palpable.

I knew I was barely scraping the surface at the Jatanyacu - quite literally. When I peered into the crevices after digging them out, I could see so much material left behind that I couldn't reach. Gold being heavier than sand meant I was missing out on the really good stuff.

I'd brought a 12 volt crack-vac, a kind of mini-dredge, down from the States, but had stored it in Vilcabamba when I'd left for the coast. I thought I'd be digging in burial mounds not crevicing river-rock, and it was too heavy a piece of kit to carry around if I wasn't planning on using it. But now I needed it. Vacuuming those cracks would be like printing money.

Once I began thinking about what could be done with the crack-vac, it didn't take long to move on to what could be done with two or three or more. Before long, I'd done the math and formulated a plan. When I found the perfect hidden cove on a bend in the river, I called Ravi in Vilcabamba and explained what I had in mind.

"Count me in," he said immediately. "Ezra will be thrilled to have an adventure, too. It'll be very educational for him. I'm sure Dana would love a vacation from the village as well."

"It'll be hard work, Ravi. I'm not sure bringing the whole family will be a good idea."

"They'll be fine. How many bodies will you need?"

"Maybe six guys, including us. Plus Dana and Ezra, of course. So, eight in total."

"Okay, I'll ask around and see what I can come up with. You really think there's enough gold there to share between so many people?"

A sudden panic gripped me. There was plenty of gold, but did I really want to share my lucrative secrets with strangers? I trusted Ravi but was uncertain about his judge of character. I thought of all the oddballs in Curly's and the juice bar, and my stomach tightened. "Tread carefully, Ravi," I said. "Only ask people you can trust, and even then don't tell them exactly where we'll be working."

Ravi laughed. "Don't worry, my friend. I've forgotten the name of the place already. Somewhere in the Amazon is all I'll say, and I won't even say that much to most."

On my way back to Vilcabamba, I stopped in a city called Puyo where I found a store selling supplies for prospectors. I ordered another *Gold Cube*, for washing the gold from the sand, and another crack-vac. I also put some pans and other gear on hold to pick up when we came back through.

In Vilcabamba, Ravi met me from the bus station in a taxi from the local cooperative. I knew the driver, a Vilcabamba native called Gabriel who was popular with the expats due to his honesty, reliability and fluency in English. On the way to his house, Ravi told me Gabriel had volunteered to join us as crew leader and translator.

"I thought a crew of local Ecuadorians would be better," Ravi explained. "They'll work harder and be more useful than foreigners."

"Sounds good to me." I smiled and clapped Gabriel on the back. "Welcome aboard. Glad you're coming with us."

It seemed I didn't have to worry. Ravi had just proven his wisdom and character judgment were faultless. One of the other guys Ravi had picked happened to be a taxi driver, too. Between him and Gabriel, we even had transportation covered. Everything was falling into place.

The twelve-hour drive to Tena was a lot of fun. After being alone for so long, it was good to be among friends again. All of us were excited about the expedition, and Gabriel kept making jokes about how rich we were going to be. We stopped to pick up the extra supplies in Puyo, then drove the final stretch to Tena, where we booked into the hostel where I usually stayed; a small family run place called Las Orquídeas.

While everyone else settled in, Ravi, Gabriel and I rushed over to the office of the local indigenous cooperative, whom we'd been told had the final say on who could prospect for gold in the Llanganates National Park. I'd been operating under the radar until now, but as we were about to take things up a notch I figured it was time to get official, or at least as official as things got in the Ecuadorian boonies.

The president of the cooperative, a cross-eyed man with a wispy mustache, welcomed us in and explained, through Gabriel, that a permit to prospect within the park would cost two hundred dollars a year. I paid him and in return received a hand-written note with the cooperative's official stamp on it.

We now had authorized clearance to mine the river, and it'd cost less than I'd made in a day when I'd been in the park by myself. I almost felt bad for not getting the permission earlier.

Early in the morning, we left Dana and Ezra to explore Tena while the rest of us drove out to the Jatanyacu and began packing in our supplies. Aside from the prospecting equipment, we had seven days of food provisions, tents, cookware and other sundries to haul along the trail to our campsite down at the river's edge. The process took hours, but by mid-afternoon we had everything set up.

Over a late lunch of canned tuna, lentils and rice, I explained how the operation would work and gave everyone a crash course in gold panning. I set up Inti, Juan and Martin, our three other crew members, to dig and pan test holes on the beach, in order to locate promising seams, while Ravi and I began crevicing with the crack-vacs. Gabriel manned a *Gold Cube,* to wash the gold from the material we were pulling out.

Once the guys identified a decent seam, I planned to pull two of them off 'exploratory' and get them to dig the seam and sluice everything through the other *Gold Cube.* The remaining man would continue digging test holes in order to find the next seam for us to exploit.

It was simple prospecting, but with the amount of gold I knew to be there, we'd soon be pulling out quite a considerable amount every day. Providing everybody was willing to work hard for it.

Aside from struggling at first with the concept of digging test holes in a grid pattern, the Ecuadorians were superb. Even Gabriel and Juan, who made their living sitting on their ass driving taxis, proved to be no slouches. Ravi was the only one not physically up to the challenge to begin with, but I knew with his drive and enthusiasm he'd soon be working at the same pace as the rest of us.

I missed the peace of prospecting alone, but the camaraderie we were developing as a group more than made up for it. All was going well, until a crowd of local indigenous appeared on the beach. They shouted at us to stop and come talk to them. I told Ravi and the others to keep going while Gabriel and I found out what they wanted, but the locals insisted we all put down our tools and hear them out.

There were several grim-faced men, some older, some younger, a few young women, a sprinkling of sullen children, and a barrel-shaped woman of middle age who appeared to be the leader. She and Gabriel commenced a heated discussion while the rest of us looked on bemused.

After a while Gabriel turned to me and said, "Do you have the permission from the cooperative? This lady says all the land this side of the river is hers. She says it's private land. Like a buffer zone before the government park."

I pulled the crumpled paper from my shirt pocket and handed it to the woman. She glanced at it then screwed it up and threw it on the ground at my feet, and she and Gabriel recommenced their rattling argument. While they shouted back and forth, I retrieved the permission slip and flattened it out against my thigh.

"She says the cooperative is nothing to her and we have to leave," said Gabriel, shrugging. "Unless we offer to pay something is what she's really saying, of course."

"I'm not paying again," I insisted. "We already paid her cousin in town. This is just a shake-down."

"Probably," agreed Gabriel. "I asked to see her land title papers and one of her sons ran off to go get them. Do you want to wait until he comes back?"

I looked up at the sun. It was already getting late and we'd barely started. I didn't want to lose any more time. "How much does she want?"

The woman blurted something and Gabriel turned to me. "Forty dollars a week. I could probably haggle her down to thirty."

"Here," said Ravi, pushing past to hand Gabriel two twenties. "That's for this week. If we do well enough tell her we'll pay her the same next week. Adiós, señora. Gracias."

The mood of the group changed the moment the money was handed over. They dropped the surliness and begun shaking our hands and asking questions about where we were from. We exchanged a few pleasantries then made our excuses and returned to work. Our new friends waved and smiled at us as they left the beach.

Half an hour later another indigenous woman appeared. This one stood watching us in silence. She looked to be about the same age as the owner, but whereas the other woman had been plump and stocky and energetic, this one was rail-thin and somber.

Juan, who was digging closest to the woman, greeted her, and after a brief exchange, he came over to where Ravi, Gabriel and I were working. "New problem," he said with a grin. "This lady says that she is owner and wants that we go."

"You've got to be kidding me," said Ravi. "How much does this one want?"

"She says doesn't want 'dirty gringo money'," said Juan. "I ask already. She just wants that we go."

"Let's go find out what's happening," said Gabriel.

This time the conversation was brief. The woman's accent was thick and whispery, but her words were firm. She wasn't making a show or wheedling for money, like the other 'owner' had.

Gabriel rubbed at the stubble on his cheek. "Juan's right, she says she's the real owner and she wants us to leave. She heard the other lady boasting about taking our money, but she doesn't want anything from us. I think she's a bit crazy."

I looked to Ravi. "What do you think we should do?"

"Well, we've already paid twice to be here. It's good she doesn't want anything, but tell her we can't stop now."

I nodded in agreement.

Gabriel translated, explaining our point of view. The woman listened attentively, glancing every now and then at Ravi and I. She pointed to our boots then back at the river, muttered something and walked away, without looking back.

"I'm guessing she asked us to go jump in the river?" said Ravi.

Gabriel shook his head and laughed. "Crazy woman. She doesn't make sense. Don't worry about it."

I was curious to know what the woman had actually said, but figured we could talk about it later. We were all anxious to get back to work. Soon after, however, we had our third and final interruption of the day: a rainstorm.

It was still the dry season, and from what I understood, the Jatanyacu was now at a record low, meaning we had a lot more rocks and sand available to scour for gold. With this in mind, alongside the fact we'd have to stop working until it passed, the rain was not welcome at all. We gathered our things and rushed to take shelter beneath the trees where we'd made our camp, stashing the full buckets of gravel we'd yet to wash on a high rock nearby on the way.

Tropical rain is intense, even when it's only a shower. It comes on fast and hard, soaking everything in seconds. Standing at the edge of our camp, forty feet up and a hundred feet back from our worksite, I couldn't even see the beach. The deluge was so intense it was like trying to look through a waterfall in the black of night.

"Amazon bloody rainforest!" Ravi shouted, laughing and ringing out his shirt. "This is normal, right?"

"Yeah, it's to be expected. I just hope the river doesn't swell too much. If it does we'll have to start our test grid from scratch again tomorrow."

Ravi clapped me on the back. "Tomorrow we can start early. It'll be a better day, I'm sure."

We ate a cold dinner of granola bars and fruits, washed down with some Havana rum to keep out the damp. As it seemed unlikely the rain would stop before nightfall, we retired to our tents to get some rest. I read for a while, listening to the rain on the tarp overhead. Every so often I'd stop to count the seconds between lightning flashes and thunderclaps. One second equals one mile, or so I'd been told as a boy. When night fell and the storm was still going strong, I shimmied into my sleeping bag and closed my eyes.

The racket of the storm kept crashing my thoughts as I drifted toward sleep. In my dream a dump truck tipped a load of stones next to my tent and I jumped up, eyes springing open. The tent lit up, bright and white and vivid, then fell back into darkness. I shook my head. There weren't any dump trucks in the jungle; it was just the storm. I laid down again and closed my eyes. The thunder kept going, ceaseless, out of time, more frequent than the lightning. It made no sense. I heard voices outside. Someone was shouting. I sat back up, awake and fully alert now. My tent shook.

"Hola! Hola! Peligro! El rio está aquí!" somebody shouted from outside.

I pulled on my t-shirt and boots and unzipped the tent. It was pitch black and the rain pelted down. The constant crashing thunder was discombobulating, surreal and terrifying.

Lightning flashed, and I saw Gabriel and Juan frozen in motion outside Ravi's tent, Inti gathering things from the ground. There was something else, too. A movement, no, a *motion* nearby that shouldn't have been there. The sky lit up again and my fear was confirmed; the river, dark and wild, was almost at our camp.

"Shit!"

I pulled on my headlamp and joined the others. They were huddled outside Ravi's tent now, flashlights in hands and on heads. Blinding beams of light cut through the downpour. I caught sight of Ravi, eyes wide and searching.

"What should we do?" I asked.

His headlamp turned on me and I squinted and looked away. "Is there any reason we should think the water level will drop?" he shouted.

It took me a moment to process his question because, in that moment, I saw a boulder fly past in the flow of the river. Another booming clap of thunder and water sprayed over us in a wave. It wasn't thunder I'd been hearing, it was boulders smashing down the river.

I shook my head at Ravi. "Shit, no! Let's pack up!"

Our Ecuadorian friends set to work, while Ravi and I remained frantic and ineffective. He gripped my arm. "Where can we go? There's nowhere! What are we going to do with all our stuff? We're going to lose everything! Jason, the buckets! What about the buckets?"

He meant the buckets full of raw material we'd pulled from the crevices. I took a step toward the raging river. The buckets were still perched on the high rock where we'd stowed them. Only the rock didn't seem quite so high anymore. It was pressed up against a massive boulder. I remembered seeing the high-water line half way up the boulder, and a few feet above the rock, earlier.

"I'll get them," I shouted over the din of clashing boulders, roaring river, rain and thunder.

"Okay, okay," said Ravi. "I'll fetch my stuff from the tent."

Reassured about the buckets, and with his mind focused on something to do, Ravi appeared slightly less panicked. This made me feel better, if only for a moment; I still had to rescue our day's work, which in the chaos of the storm, I saw as my side of the bargain to help keep him calm.

The water was ankle deep, the current tugging, at the foot of the rock. I climbed up on top and grabbed one of the buckets. It was heavy. At least seventy pounds. Gravel and rainwater sloshed as I lifted it to chest height and scrabbled up the steep, slick slant of the higher boulder.

My feet slipped and my knees smacked against rock. Rain beat down into my eyes. I couldn't see and I couldn't hear anything beyond the sounds of nature's fury. Disorientated and scared out of my wits, I inched my way up the boulder's edge. Eventually, I reached the top and plunked the bucket down.

I turned to go back for the second bucket but lost my footing and tumbled, sliding downward, out of control. I was sure I'd end up in the river, but by some miracle I slammed into the rock. Struggling to my knees and gasping for breath, my elbow hit against something smooth. I jumped and almost fell again. Squinting through the rain to see what it was, I saw the remaining bucket.

In spite of my survival instincts screaming for me to leave it and get to the safety of the bank, I couldn't; I was caught in the manic grip of my insane quest. I grabbed the bucket and began the suicidal trek up the boulder once again. When I got to the top, I placed it next to the first and backed away carefully. I couldn't count on a second miracle saving me.

I lowered my head side-on to the rock face. The beam of my headlamp picked out what looked like a rushing stream but was actually my path down the boulder to the rock below. Orientating myself as best I could, I let go and skidded down, my feet, hands and butt scuffing up spray all the way. Only at the last moment did I see I was veering away from my target. The rock I'd been aiming for flashed past, and I fell headlong into the water.

My only comfort in the split-second after I left the rock-face was at least I was on the shore side where it was shallow. But when I splashed down it wasn't shallow at all. The river was deeper and the current a lot stronger than it had been on my way out. I struggled to plant my feet into the shifting sand and raise my head above the surface. Gasping, I lurched toward the rock and hugged it. The water level was at my chest when it'd been ankle deep only minutes before. The river was rising at a rate I hadn't imagined possible. Pushing off the rock, I swam hard for the shore. With the tug of the current pulling me further and further downstream, I only just made it.

Fortunately, my headlamp was still working and I was able to trek back through the trees and bushes to our camp. When I arrived, I had another shock. It was gone. Tents, hammocks, pots, pans, crack-vacs, supplies; everything had disappeared. Only a weak, flickering light remained. The light moved toward me. It was Juan.

"Señor Jason, follow me," he said, dipping the beam of his flashlight toward the ground.

We scrambled up the tumble of white water that used to be our dry, creek-bed trail. When we reached the main track through the jungle, Juan turned left and walked a few paces before turning right along a half-hidden trail through dense bushes. We burst into a wide clearing where we found the others in a bustle of activity, remaking our camp. Most of the tents were up already, but there was still plenty to do. Juan and I joined them and we soon had everything stashed and tied in place. We bid each other goodnight with a round of hugs and slaps on the back and went back to our damp tents. Incredibly, I fell into a deep sleep within seconds of lying down.

During the night, the storm subsided, until only the patter of light raindrops remained when I awoke the next morning. Outside, the world was leaden-gray and luminous green. The sky and jungle misted and blurred into each other; *this* was the rainforest! Everyone else was up already. They were clustered together, drinking coffee and complaining beneath a mix of wool and plastic ponchos.

"We want to go home," said Gabriel, handing me a coffee mug.

The other Ecuadorians nodded. Ravi stared at the ground.

"Ravi?"

He looked up, his expression blank. "I need to get back to Tena. To see Dana and Ezra. They would have heard about the storm. They'll be worried."

"Okay, guys," I said. "Let's pack up and go back to town. We can figure out what to do when we get there. We don't have to make any rash decisions."

It took a long time to carry everything back along the squelching trails. The excitement that had buoyed us along the previous day had dissolved into soggy dejection. Nobody wanted to go anywhere near the river's edge, and we all took extra care crossing the rope bridge out to the muddy parking lot.

While the others packed our stuff into the pick-ups, I made a lone pilgrimage back down to our camp. The river had receded but judging by the damage to the trees and foliage and the gouged-out topography, it was pretty clear we'd have been washed away if we hadn't escaped in the night. To my amazement, however, the two buckets I'd rescued were still sat proudly on top of the boulder. I hauled them down, and carried them back, filled with a morbid curiosity to know how much gold had almost cost me my life.

When we arrived in Tena, the town was in a terrible state. It had been the worst storm to hit the region in fifty years and the damage could be seen everywhere. It was like a war zone. The suspension bridge across the River Tena had been mangled, and many homes were washed away.

We found Dana and Ezra holed up at the hostel, sick with worry and fright. They and Ravi had such a joyful, tearful reunion it made my heart ache to watch. My conscience pricked with the guilt of what could have been had we not made it out safely.

In the evening, I washed the gold out of the gravel in the buckets. There was less than two grams. I'd risked all our lives for next-to-nothing. The next day we left Tena and went back to Vilcabamba.

A few days later, I saw Gabriel outside Curly's bar. We talked about the trip, and I finally got to ask him what the crazy old lady had said.

"You know, Jason, I don't think she's crazy anymore."

"Why? Because we turned out to be even crazier?"

"Ha! Maybe. But what I mean is she told me, *us*, to be careful. She knew the storm was coming, although she didn't say it outright."

"What did she say exactly?"

I knew the answer, or at least a part of it, before it left his lips.

"She said the river was hungry. And if we didn't leave it would gobble us all up." He made the sign of the cross. "We were lucky to have survived. I'm never going back there, Jason. Never."

I understood how he felt, but he didn't know the Jatanyacu like I did. I was already thinking about how much gold must have been washed in during the flood and regretting not staying to pull it out.

Huaorani Prince

After our narrow escape from the Jatanyacu, I went back to working by myself. Risking my own life was one thing, but dragging family men like Ravi and Gabriel into my schemes was a responsibility I could do without. I also preferred the peace of working alone to the stress of organizing and running a large crew.

Although I was yet to meet Fred Miller in person, he kept coming through with great leads. On his advice, I worked my way around Ecuador's gold hotspots, from El Oro province in the south, to Esmeraldas in the north. I did well financially and amassed a wealth of local knowledge along the way. So much so, I became a new point of reference for Ecuador on the Internet treasure forums.

On one of these forums I was introduced to Tyaento, a Huaorani tribesman, whose mother was made famous in various movies and documentaries about the killing of five Christian missionaries in Huaorani territory back in 1956. One of these movies, *End of the Spear*, I'd watched while preparing to leave for Ecuador.

It told the story of how the Huaorani – who were also known as Aucas, the Quichwa word for savages - resisted all outside contact until just after the massacre, when Tyaento's mother, a Huaorani herself, took the families of the dead missionaries to live with her people. Tyaento's mother, who was a young woman at the time, had fled the jungle as a child, and had to win back the trust of the Huaorani before she and the foreign women and children were allowed to stay. This she did, and in the end many Huaorani subsequently renounced violence and converted to Christianity.

Now, more than fifty years later, her son wanted me to come into their territory and teach the Huaorani how to dredge their rivers for gold, in exchange for a half-share of everything we could pull out while there. I knew the river system Tyaento described was a sure bet, and I was intrigued to see how much this infamous tribe had changed since the nineteen-fifties. I agreed to the deal without hesitation.

First though, I needed to put together a team of three guys and find a dredge. Although I had a lot of dredging experience, I was not an expert, especially when it came to the engineering know-how. So I put out a call on the forums for a professional dredger and someone with good mechanical and survival skills. I also began making enquiries about acquiring a dredge locally.

The first person to get in touch was my old friend Henry. He told me the reason he'd not been back to explore Coaque yet was because he'd been up north working a lucrative dredging contract. He was finishing up for the winter and offered to come down and build the dredge from a combination of local parts and specialist equipment he could bring with him. I couldn't have been happier. Henry and I got along well, and he was also tough and tenacious enough to handle life in the jungle.

The third member of our team was Simon, a survivalist and off-grid living expert from Nevada. He was looking for adventure, had all the right skills and was already in the country. The only downside was he was travelling in Ecuador with his girlfriend, a Latvian lady called Iveta. He told us Iveta was used to harsh conditions and would be willing to help out for free if we agreed to her coming along. So, our team of three became four.

We all met up in Vilcabamba six weeks before we were due to head out to the Huaorani territory. I wrote to Tyaento to see if he'd come meet the team and finalize all the arrangements so we could get started on building the dredge. He was in Cuenca, a city four hours north of Vilcabamba, visiting family, and said he'd come down for a few days if I gave him two-hundred dollars to cover expenses.

"I'm just the translator," he insisted when we spoke on the phone. Both his written and spoken English were perfect. The missionaries with whom his mother worked had sent him to be educated in the States. "The dredging deal is with my sister and brother-in-law. I'm not making anything. I'm doing it all for my people and for you guys. I don't want to be left out of pocket, though. Buses and hotels and food cost money. So, if you want me to come talk with you all there you'll have to help me out." I agreed to pay the two-hundred bucks and the next day he arrived in town, with his wife.

Tyaento was in his late forties or early fifties. He was handsome, and very slim and tall for an Ecuadorian indigenous man. With his long, graying hair swept back into a braid, blue jeans and a red plaid cotton shirt, he looked more like a Native American of the north. His wife was a frumpy white-haired American, decked out in pastel shades and jungle beads.

"You must be Jason," she said to Henry, when we all met up in

Curly's Bar. "You're exactly how I imagined!"

"I'm flattered," he said, smiling. "But I'm Henry, the engineer." He nodded at me. "He's the adventurer."

She looked me up and down, unconvinced. "Okay. Well, I'm June, and this is Tyaento, Prince of the Huaorani."

We all looked at Tyaento. He wasn't paying attention to the introductions; instead, he was shamelessly ogling Iveta's long bare legs. Simon put his arm around Iveta's shoulders and June clicked her fingers in front of Tyaento's nose. He grunted, then smiled and gave each of us a firm handshake.

We pulled two tables together, ordered food and beers, then set to work discussing the expedition.

"Have any of you been in the Amazon before?" Tyaento asked.

Simon and I nodded. Iveta and Henry shook their heads.

"I mean really into the jungle, not the outskirts with all the other tourists. Where we're going you'll see jaguars, anacondas, monkeys. Things you've never seen before. Huaorani land is magical. It's the real jungle. Pristine and beautiful."

"It really is!" enthused June, through a mouthful of pizza. "Tell them about it, love."

Tyaento nodded. "But it can be dangerous, too. You'll be far away from all the comforts and security of western civilization, living with the Huaorani for the duration of your visit."

"Sounds perfect," I said.

"We'll be fine," said Simon. "I've had training and we'll take in the things we'll need to stay safe."

Tyaento raised his eyebrows. "What do you have in mind? We don't want guns brought in to our territory."

Simon chuckled. "I wouldn't know where to find a gun around here even if I wanted one. I meant medicines, tools, that kind of thing. Maybe you can help me put together a list?"

"Of course," said Tyaento, breaking into a lazy smile. "I'm here to help you guys. To be a bridge between you and my people, and our land."

"Speaking of which, can you tell us what you and the Huaorani expect from us in return for letting us onto your land?" Henry asked. "Jason's explained as much as he knows, but it would be good to have everything open and clear from the start."

"Sure. Like I said, I'm only the facilitator. The deal you have is with my sister and her husband. The land is theirs, and you'll be teaching my brother-in-law and two nephews how to use the dredge. You do have the dredge, right?"

"We're building it," said Henry. "Importing a whole dredge makes no sense. I brought in the stuff we can't get here. The rest shouldn't take too long to put together."

"We were planning on building it in Tena. Then we'll bring it down to you guys," I added.

"Sounds good. If you want, you can all stay with June and I in Puyo while the dredge is being built. There's plenty of space and it'll save you money on hostels."

Everyone was smiling and nodding; this was going to be good.

"Thanks, Tyaento. We'll do that."

"Cool. So, when we get to my sister's place all the gold we find will be split fifty-fifty. Half for the gringos," he gestured to us, as if there might be some confusion as to whom he meant, "half to the Huaorani. Once our people know how to use the dredge, perhaps you can show us how to make one so we can have our own?"

"Absolutely," said Henry. "We'll introduce you to the metalworkers in Tena, and I could bring the rest of the kit down for you next time I fly in."

"How big is the settlement?" interjected Simon. "I'm trying to picture it. How many people live there?"

"It's small. Only ten or so people, but we'll make sure you're comfortable." Here he nodded to Iveta. "There's an old schoolhouse left by the Summer Institute where you can all stay." Tyaento then turned to me. "In exchange for accommodation would you supply the food for the men?"

"The men?"

"Everyone working the dredge. You guys, me, my brother-in-law and nephews."

"Sure. Seems reasonable."

"Would you like us to help with some other stuff while we're there?" asked Simon. Tyaento smiled his languid smile and waited for Simon to continue. "I mean - do you have bathrooms?" he asked. Tyaento grinned and shook his head. "We could build you some composting toilets. Also, Jason could help you with your crops. He's a permaculture expert. How would that be? We'd love to help out and do some projects during our free time if you're interested."

Tyaento nodded his assent. "Thanks for thinking of it."

With everything ironed out, we spent the rest of the evening laughing and drinking beers. Tyaento told us lots of jungle stories and by the end of the evening we all felt like firm friends. When Curly came over to tell us he'd be closing up soon, Tyaento pulled me to one side.

"Do you have the cash for our expenses?" he asked. "It's just the hostel wants us to pay up front for the three nights. I already put them off once, but I'll have to give them the money in the morning." Curly leaned over at this moment and placed our bill on the table between Tyaento and I. Tyaento glanced at it. "And I want to pay our share for tonight," he added.

Feeling good, I told him not to worry - tonight was on me. I then passed him the two-hundred dollars. He looked at the money and frowned.

"What's the matter? Didn't we agree on two?"

"Well, yes, but what about my wife? She'll need two as well."

I laughed. "Are you serious? She didn't have to come. Besides, she's an American. Can't she pay her own way?"

Tyaento laughed with me. "I guess so. Don't worry about it. Two's fine, my friend."

Apart from the minor awkwardness over Tyaento's expenses, I felt the evening had gone well. Henry, Simon and Iveta agreed. We were even more excited about the trip now. Meeting Tyaento had made it all so much more real. He was such a cool and entertaining guy.

The next few days were spent talking about the finer details, and getting to know each other better. When Tyaento and June left to go back to Cuenca, everything was set and we were all ready for the adventure.

We had a few weeks to plan and dream some more while waiting for the dry season to arrive in the eastern jungle. Henry and I collected materials for the dredge and spent endless hours discussing where we would go with it once we'd finished training the Huaorani. Knowing how hard conditions were going to be in the jungle, Simon and Iveta went to the coast to relax until it was time for us to meet up again.

When Tyaento called to say the heavy rains had ended, Henry and I packed up our things and headed straight to Puyo. Tyaento and June lived in an airy two-story house with five bedrooms. They made us welcome and showed us around town the first day, including a visit to the shitty burger bar where June loved to eat because it reminded her of home.

The next day, Henry and I went to Tena to meet with the fabricators who were making parts for the dredge. Everything went well. We were able to collect some of the parts and made arrangements to pick up the rest a few days later. That night, Simon and Iveta arrived, tanned and cheerful, and we had a party to celebrate our reunion and the real beginning of our adventure.

The party went on until way past midnight. Tyaento was in great form, until a friend of his came by to drop off a huge brick of hashish. A couple of hefty joints later and Tyaento fell asleep while describing the complex ties between the missionaries who raised him, their corrupt language institute, the American oil companies and the C.I.A.

The next morning, we were sitting on the porch when June brought out some cold sodas. "Have you seen Tyaento?" she asked, placing the tray on a low table and sitting down beside Henry.

We told her we hadn't.

"That's because they're avoiding me," Tyaento said, emerging from behind an unkempt bougainvillea. "They didn't like being told what your people have done to mine."

We were all disturbed by Tyaento's sudden appearance. Not to mention his aggressive attitude. I wondered what he'd been doing in the garden. He had no tools to hand.

Henry was the first to regain his composure. "Actually, I found it both fascinating and depressing," he said. "I only wished you'd stayed awake to finish the story. How long have you been hiding in that bush, by the way?"

"Long enough. I didn't finish because the story is not yet done. Why are you all lazing around and not working on the dredge? We need to get it out to the land."

"Like we told you yesterday, we have to wait for the rest of the parts. Why are you stressing? We've got all week."

"Do we? Typical gringos. Always taking advantage of Huaorani people. You know what you are? You're fucking termites, that's what!"

"Tyaento, darling, have a coke and cool down. Please." June picked up a glass and held it out to him.

"Don't think that doesn't include you, woman," he said, snatching the glass from her hand and stomping off indoors.

"Hmmm, sorry about that," said June, standing and smoothing the creases from her skirt. "The Huaorani *are* natural born warriors after all. Enjoy the cokes and don't fret about him. He'll soon cool down." With that, she set off into the house as well.

When she was gone, Simon let out a long, low whistle.

Iveta frowned and said, "Do you suppose they worry so much because it's dangerous and there's not so much food where they come from?"

"What? Who?" I asked.

"The Huaorani."

Henry chuckled. "I think she said warriors not worriers. Though you could be forgiven for thinking it was the other way around. Jeez, what's got into that guy?"

I shrugged. "Who knows what he's dealing with. He must be stressed about something."

Simon drained his coke and belched loudly. "Well, whatever it is, I'd advise caution from here on. Check behind the shower curtain when entering the bathroom and behind every flowerpot when stepping into the garden. Who knows where our gracious host might be lurking, waiting to pounce."

We all laughed, if a little uneasily. I glanced at the open front door; many a true word is said in jest.

*

The fabricators in Tena got some of the sluice dimensions wrong, which resulted in a two-day delay. Tyaento, who'd been distant and sulky since the morning after the party, was now furious.

"What's wrong with you gringos? Can't you get anything right?" he hissed at Henry when we told him about the complications.

"Actually we can, and we are, which is why we need the extra time," Henry explained. "It's the Ecuadorians who messed up, not us. They're going to remake the sluices for free, but it'll take a couple more days, that's all."

"That's all? Easy for you to say! Your incompetence is going to make me look a fool and a liar."

I could sense he was about to storm off again so I stepped in front of him. His eyes narrowed and I felt the hairs on the back of my neck stand up. "Look Tyaento, we need to get this right before we leave. Otherwise we're all going to look like fools and liars. What's the point of taking a dredge that doesn't work out to your people?"

"Two more days. If it's not ready by then, you can all leave and I'll find someone else to help us." Scowling, he pushed past me and out of the room.

"Why are we helping this guy again?" Henry asked.

"It'll be fine once we get out there," I said, trying to reassure us both. "I'm sure he's just nervous and feeling the pressure."

We managed to get the dredge put together in time so we never found out if Tyaento's threat had been idle or not. Once everything was ready, his mood lightened and he became more like the laid-back guy we'd met in Vilcabamba.

"I've sent word to my people. We'll have help packing everything in tomorrow morning," he said, rubbing his hands together. "I have to admit, for a moment there I thought you guys were chancers. But it's all coming together. Good work guys!"

"Thanks," said Henry, smiling in spite of himself.

"Hey Tyaento, does your sister have any little kids?" I asked.

"Of course," he grinned. "We all do! Why?"

"I'm trying to work out how much chocolate and candy to buy. We'll need some for energy, but I thought it might be nice to give some treats to the kids, too."

"Good idea. Let's bring as much as we can. I'm sure it'll get eaten."

I marked down a couple of extra boxes on my list. Simon, Iveta and I went to buy the supplies while Henry stayed behind to tinker with the dredge. We bought a first-aid-kit, gardening and construction tools, batteries, and enough food to last eight people three weeks. Mostly eggs, dry goods, powdered soups, rice, protein bars and chocolate. We figured fresh fruit and vegetables would be abundant in the jungle.

In the morning, we all piled into the truck we'd hired to transport the dredge and headed southeast on the main road out of town. After a while, we turned east onto a paved road that led to a small town. We stopped for breakfast, then took a dirt road into the jungle. The trees grew older and higher, the foliage wilder, with every bumpy mile we traveled. Eventually we stopped beside a narrow river.

A huddle of bare-chested men greeted us and helped unload the dredge and other supplies. Unlike most Amazonian tribes, who seem to favor pageboy bowl-cuts, the Huaorani wore their hair shoulder length with short, severe bangs. They were all happy to see Tyaento, embracing him and clapping him on the back while pointing and smiling at the dredge and us gringos.

Tyaento introduced me to Dabo, his brother-in-law. Dabo smiled and pumped my hand while speaking in a mix of pidgin-Spanish and Huao. I looked to Tyaento for a translation.

"He's happy you are here. Now don't let us down, eh?"

I laughed. "We don't intend to. Tell him we're happy to be here and we're willing to help in any way we can."

Tyaento grunted and said a few words in Huao. Dabo frowned and turned away to help unpack the truck.

We placed most of the provisions and supplies into the canoes the Huaorani men had brought up the river. The jerry-cans of gas for the dredge were lashed together and tied to the backs of the canoes where they bobbed and turned in the flow of the river. The dredge splashed down onto its pontoon floats to a chorus of whoops and cheers. Stacks of eggs and sacks of rice and flour were piled high onto the work platform.

The Huaorani took care of everything in the water while the rest of us put on our backpacks and followed Tyaento, Dabo and two young men along the riverside trail.

"Is this the river we'll be dredging?" I asked Tyaento. I was surprised at the lack of large boulders and how small it was – only fifty feet across for the most part.

"Yes. We're close to the headwater here. Further down it gets wider, but not by much. It should be easy work. There's a lot of gold concentrated here."

"That's what I heard. It's just not how I imagined it."

Tyaento shrugged and picked up the pace.

We'd been walking for about twenty minutes when the sky darkened and the wind picked up out of nowhere. I could smell the rain before it hit and was taken instantly back to my time at the Jatanyacu.

"Shit! There's a storm coming," I said.

"There sure is," Simon agreed, looking around at the flapping leaves. "Maybe we should stop and find shelter, or go back. Storms in heavy jungle can be dangerous."

"Stop whining," Tyaento called back to us over the rushing wind and insistent grind and rustle from the trees. "Come on! We need to get there!"

The clouds opened and rain began falling in swirling sheets, turning the trail to slick mud. We slipped and dug our way forward through it. I couldn't see our supplies or the dredge out on the river anymore. Just a torrent of frothy rapids. I reached Tyaento and grabbed his arm.

"What about the river? Our supplies are out there!"

He pulled his arm from my grasp and whirled around to face me. "The river's the safest place at the moment. It's still not deep and the current will help them reach the village sooner. We're the ones in danger."

As if on cue, we heard an earsplitting crack, followed by fast-snapping branches. Tyaento gripped my shoulder and the two of us watched a huge tree limb crash into the ground ten feet ahead of us, spattering our faces in mud.

"We have to stop!" Simon shouted. "We're going to get killed."

Tyaento looked at him with contempt. "And go where?"

I glanced at the rising river. No way I was going in there. Even before the Jatanyacu, I'd always rather take my chances on dry land.

"Tyaento is right," I said. "Let's just keep moving."

Dry land soon turned out to not be an option when we reached a tributary to the main river, which, according to Tyaento, was usually a tiny bubbling stream, but was now a fifteen foot wide rushing mass of dirty water and driftwood.

"I hope the others have arrived with the dredge," said Tyaento. "This storm is bad."

"No shit, Tyaento," said Henry. "We should have gone back when we had the chance. We should have listened to Simon." Another limb, this one at least a foot in diameter, struck the trail twenty feet back. Henry jumped, shot a look at it, and asked the question on all our minds: "What the hell are we going to do?"

Tyaento barked an order at one of the younger Huaorani. Dabo placed his hand on Tyaento's forearm and began pleading with him in Huao. In the meantime the young man hoisted the sack he was carrying high up on his shoulder and began crossing the river. Tyaento and Dabo fell silent, and we all watched as the youth, who couldn't have been more than sixteen, picked his way through the waist-high churning water. Sticks and branches crashed against his body before whirling away downstream. He reached the other side and with a wide grin beckoned us all to follow.

Not to be outdone, the older Huaorani, including Tyaento, charged into the river, whooping and yapping at each other. With a little more trepidation, Henry, Simon, Iveta and I followed.

The water was warm and the current so strong it was almost impossible to find a foothold. I stumbled and splashed and leaned into the flow, bracing myself against the constant slap of debris and trying desperately not to fall.

Now we were out of the trees, the full force of the sixty-mile-an-hour winds buffeted my backpack and upper body. The rain hit my face sideways, coming from the opposite direction to the undercurrent. The combination gave the feeling of being attacked from both sides. Any false move and I'd spin away, head over heels, and be consumed by the roiling river.

At one point I stumbled and lost my balance, but Iveta grabbed my shirt before I toppled over. She was stronger than she looked. Stronger than me, that's for sure. She pushed me toward Simon who was already on the bank. He reached out with both hands and hauled the two of us out of the water.

"Thanks," I gasped.

Iveta smiled demurely. Simon nodded, then lurched forward again to grab Henry. We were all safely across.

We rejoined the path, which was now a muddy stream, and tramped on. The trees creaked and groaned, and every now and then we'd hear, and sometimes see, another thick limb thump to the ground. As quickly as it came on, the storm subsided. Leaving the air calm, yet charged and muggy.

Four hours after entering the forest we crested an incline and caught our first glimpse of the Huaorani village. Children ran out to meet us, barefoot, grimy, and happy as larks.

"How many kids do you count?" Henry asked.

"Fifteen, I guess. Maybe more. Why?"

"Didn't Tyaento say there were only ten people living here? In this pristine jungle paradise?" The sarcasm in Henry's tone was hard to miss. So was the overpopulated, rundown shantytown before us.

I'd imagined a single wattle and thatch longhouse in a small clearing, but this was a bustling village. There were thirty or more wooden buildings scattered around, all in various states of decay. Trash tumbled in gusts, settling to fester in dank corners. There was even a soccer field which doubled as an open-air lavatory for dogs and children alike.

Simon nodded to a ten-year-old boy who was waving cheerfully at us while defecating between the goal posts. "Might be a good spot to start the veggie garden, Jason. Plenty of manure."

Dabo led us to a rickety platform with a wide covered porch and four doors, each leading to a separate, windowless room. A ragged line of Huaorani men and women were carrying our supplies and tools up from the river and into one of the rooms.

"You and Henry have a room each and the lovebirds will share," said Tyaento, appearing on the platform behind us.

Children were buzzing around his legs, and every second or two someone new would call out to him in greeting. Whatever we thought of Tyaento, in this village he was indeed a prince.

I opened the door to the room closest to me. The stench of rotting fruits and feces wafted out. I gagged and backed away, trying not to throw up. Tyaento laughed and wandered off to visit with his friends and family.

"Good job we brought tents," said Henry.

I nodded in agreement, thinking I might vomit if I dared open my mouth. Once I pulled the door shut, I was finally able to speak. "Let's set the tents up out here on the platform. Man, I'm exhausted."

"Me too. Not to mention hungry and soaked through."

"Yeah, all that and pissed with Tyaento."

"Don't say that too loud," said Simon, joining us. "We have to keep him sweet. He's the only one here we can communicate with."

My heart dropped at the thought. We were stranded in Huaorani territory at the mercy of a fickle, pouting prince, with no one to blame but ourselves for being there. Nothing to do now but make the best of it. Henry and I pitched our tents while Simon and Iveta cleaned out the least noxious of the rooms. Then we organized the storeroom together and set about making dinner.

While we were cooking, a couple of Huaorani came over and took some rice, lentils, yucca, and canned tomatoes from the storeroom. The thieves were so brazen, we were all too dumbfounded to say a word.

Tyaento arrived shortly afterward and asked where we'd hidden the chocolate. He was a little unsteady on his feet and his eyes were bloodshot and unfocused. It wasn't even dark yet and he was already drunk. I retrieved a box of Hershey bars, opened it and went to hand him one. He pushed my hand aside and took the box.

"For the party," he muttered, snatching the bar I'd offered him as well. He unwrapped it one handed and pushed the whole thing into his mouth, letting the wrapper flutter to the floor. "Don't forget we're on the equator," he said through the mouthful of chocolate, some of which was now dribbling down his chin. "Which means only twelve hours of daylight. So we need to start early tomorrow. Six o'clock. Make sure you're up!" He then turned and staggered away toward the mass of Huaorani gathered around a bonfire cooking our food thirty feet away.

As tired as I was, I found it impossible to sleep. Although the jungle comes alive at night, I couldn't hear any of the natural sounds over the raucous laughter, constants shouts, and the din of the tinny transistor radio playing techno-cumbia music until the early hours of the morning.

When I crawled out of my tent at five-thirty, Dabo and the two young men who'd walked to the village with us were waiting. Crouched on their haunches at the end of the porch.

"Buen día," said Dabo.

"Buenos días," I replied. "Good party?"

They didn't reply.

I woke Henry and Simon and the six of us made our way to the river in the dark. We found the dredge tied up to a tree and in need of some attention: some of the components were loose and rattling, others had been pulled apart.

"How did this happen?" asked Henry, holding up a detached hose.

Dabo smiled and nodded.

"For god's sake, where's Tyaento?"

Dabo smiled and nodded again.

"How are we supposed to communicate without our translator?" Henry said, tossing the hose over the sluices.

"Not very well by the looks of things," said Simon.

I laughed at the craziness of it all. "It's okay," I said. "We can show them what to do. And they do speak Spanish. If you two want to see what we need to do to get the dredge going, I'll start them off digging for seams."

Dabo and his sons turned out to be excellent students and hard workers. Unlike the guys from Vilcabamba, they grasped the idea of making a test grid immediately in spite of the language barrier. They also knew how to pan already.

By midday we had the dredge running and had identified a few promising seams. Having the dredge working and having it calibrated for the task at hand were two different things, however, and we struggled to get any kind of decent productivity from it.

"Don't worry, I've got it," said Henry, and he set to work on adjusting and rectifying the sluicing problems with quiet efficiency. The rest of us continued to dig and pan.

Tyaento turned up around one in the afternoon. "Why aren't you dredging?" he asked. "Why the hell are you using pans? This isn't the dark ages!"

"If you'd been here earlier you would know. Today's the first time we've tried the dredge. It's normal for it to need some adjustment in the beginning. There are so many variables to consider with a new site. Besides, we didn't find it in the best of conditions," Henry explained.

"What did you expect? It came in on a storm. I still don't understand why they're panning when we have a dredge. Shouldn't they be helping you fix it?"

"We're panning to find the richest gold deposits so we can dredge them later," I said. "It's called efficiency, Tyaento. Henry's the dredging engineer. We're letting him do his job while we do ours."

Tyaento shook his head and turned to Dabo. They spoke for a moment in Huao and when they finished Tyaento pointed an accusing finger at me. "Dabo said they've been digging holes and panning all morning. You haven't even shown them how to use the dredge yet!"

"That's because we haven't had a translator to explain anything to them," I snapped back at him.

"Always excuses with you termites! Tell me how it works and I'll explain it to them now. Have I got to do everything for you? You want me to chew your food as well, you entitled little gringo shits?"

I looked at Henry. He shrugged and said, "Okay, guys. Come over here and let's get you started."

Tyaento lasted ten minutes before he got bored and walked off toward the village. Henry, who had been in the middle of showing them how to hold the hose under the water when Tyaento up and left, sat back on his heels, shaking his head in exasperation. This was going to be a lot more difficult than we'd anticipated.

By late afternoon, Henry had the dredge working properly and we were able to give Dabo and his boys a proper demonstration. They missed a lot of the finer details but were enthusiastic to learn, and by

Falling to Fly

the end of the day we all felt a lot more positive. In spite of pulling only a miniscule amount of gold from the riverbed.

When we arrived back the village, we found the storeroom door wide open and everyone munching on our candy bars. The wrappers were strewn across the ground, drifting into the surrounding forest. Our food supplies were cooking in a big pot over the fire, and Tyaento was in the middle of a crowd shirtless Huaorani, telling a story that had them all doubled over in hysterics. Somebody handed Dabo a cup of chicha - a crude kind of beer made from fermented yucca. He drank half the cup and offered the rest to me.

"No gracias." I was too annoyed to join their party. "Tyaento, can we have a word please?"

Tyaento held up a finger, and finished his tale. At the end, the Huaorani cracked up again, holding their sides and rolling in the dirt, stealing glances at me, Henry and Simon.

"What can I do for you maestros?" Tyaento asked, sauntering over to us. "How much gold did you make for your Huaorani hosts today?"

"Never mind that. Why is the storeroom open? Fair enough you had a party yesterday, but we can't feed everybody every day."

"Are you going back on your word, gringo?"

"What? No. We're only supposed to be feeding the work crew. Besides, you said only ten people lived here."

"So? Everyone else is visiting. If it's such a big problem for you to share we can lock the storeroom. Is that what you want?"

"Yes, I think that would be a good idea." I knew I was right to put my foot down, but at the same time Tyaento was making me feel small-minded and petty.

He stalked off without another word, leaving the horde of Huaorani men and women glaring at me. Dabo and his sons looked away, and for a moment I thought we were going to be speared like the missionaries back in the fifties. Then I noticed the crucifixes many of them wore around their necks, and I felt a little safer. For better or worse, these were the tame Huaorani, not the Aucas they'd been before being 'saved' by Tyaento's mother and her missionary cronies.

Tyaento reappeared up on our sleeping platform five minutes later with a padlock and key. He secured the door and then hung the key on a nail next to it. "Happy now?" he asked.

"Yes thanks."

"So, how much gold *did* you make today?"

"Just a few grams."

"You're joking, right? You'd better not be holding back on me. Dabo is watching everything you do so you won't get away with it if you are. This isn't Vilcabamba. You can't rip me off like you did there."

"For god's sake, Tyaento, what the hell are you talking about? Nobody's trying to rip you off, and no one ever has," interrupted Henry. "Today was day one. It'll get better, much better. And anyway, don't forget who's funding this whole thing." He nodded at me. "You're not losing anything. If we don't get any gold it'll be Jason who's out of pocket, not you."

"Jason, huh?" Tyaento spat on the boards by my feet. "I'm still checking with Dabo. We'll be watching you." Before anyone could say anything else Tyaento walked away again.

"Do you think he storms off all the time because he can't bear anyone but him having the last word?" said Simon.

We all laughed at that. It had been a long day and our nerves were frayed. Tomorrow had to be better.

As far as productivity was concerned, the next day was a vast improvement. We all took turns on the various dredging jobs, including Dabo and his sons, whom we now knew were called Kimo and Moipa. We made a steady flow of gold that amounted to a little under a half-ounce by the end of the day. In the afternoon, Simon left early to start digging the composting toilets he'd promised to make for the Huaorani.

"I do hope they'll use them," he said, before heading off. "But at the very least *we* won't have to shit on the soccer field anymore."

*

At the end of day four, Dabo gave his wife all the gold we'd made so she could go to Puyo and cash it in and replenish our food supplies. Due to Tyaento keeping a key to the padlock, we were still feeding the whole village. There were less of them now, however. Tyaento had been right about one thing: over half the people who'd been there when we first arrived had been visiting from other settlements. By day five, though, more visitors had drifted in. This pattern continued right up until we left. Word of the gringos handing out free food must have spread quickly on the jungle telegraph.

We were working grueling twelve-hour days, slinging rocks, digging, panning, diving in, and often inadvertently swallowing, the murky waters of the river. Our bodies were being pushed to the limit. It was only a matter of time before our immune systems failed and we started getting sick. Henry was first. He came down with a high temperature and diarrhea, meaning Simon and I had to run the dredging for half a week without him.

Up until then, the three of us had rotated our duties in order to work on other projects for the Huaorani. Simon constructed composting toilets throughout the village with Iveta, while I set up and planted gardens. I'd wanted to teach the Huaorani about permaculture, to provide them with an alternative to the 'traditional' slash and burn farming methods so damaging to the delicate rainforest ecosystem. Unfortunately, Tyaento was usually too busy drinking chicha and smoking hashish to translate for me, so I mainly demonstrated techniques and practices without explanation. Something the children and elders who gathered to watch me found baffling, and often hilarious.

Henry was only out of action for a few days. A combination of rest, antibiotics, rehydration salts and a hardy disposition saw him back to full health quickly enough. In the meantime, though, Simon had developed a more severe problem.

After dredging all day in shorts and rubber boots with no socks, he'd managed to scour the skin on his left leg raw from the middle of his calf to his heel. The next day he wore socks and a light bandage over the bloody blisters, but the damage was already done. By day three his wounds were infected and full of pus, and he succumbed to a fever.

Iveta took care of him; bathing his leg three time a day, applying steroid creams, and administering elephant-size antibiotic pills. Simon was even bigger and more robust than Henry, so we didn't worry at first. But within a few days it was clear things were getting worse. He could hardly walk, wouldn't eat, and was delirious, slipping in and out of consciousness and babbling nonsense.

About a week after he was first injured, Iveta came to Henry and me at the end of our workday frantic with worry. Simon's lower leg was swollen and red-hot to the touch. The initial lacerations were weeping a foul-smelling, greenish discharge, and the surrounding area was a web of dark, marbling veins that reached to his knee.

"We need to get him to a hospital," said Henry. "If he doesn't get proper medical help soon he's going to lose that leg. Shit, he might even die. Let's see if we can borrow a canoe. I'll pull him up the river and out of here tomorrow morning."

"I'll go ask Tyaento," I said, ducking out of the room. The reek from Simon's leg was making me nauseous, and I was glad to have an excuse to leave.

I found Tyaento with his drinking buddies. He ignored me until I pushed through the small crowd and stood directly over his hammock. I explained how bad things had gotten with Simon, and asked for his help in arranging a canoe. He laughed in my face, and then spoke loudly in Huao to the gathered throng. I felt the mood shift, and not for the better.

Tyaento shrugged. "Nobody wants to help you termites. It's not our fault the man is weak. But if he dies from his wounds, I promise to take good care of his woman."

I wanted to punch the leering smile off his face, but knew if I did it would cause a riot. Me getting beaten up wasn't going to help Simon. "You're such an asshole!" I said, then turned and walked away. There was no way I was going to let the bastard get the last word after what he'd just said.

As I made my way back to my friends, I noticed somebody had peeled away from the group and was following me. I whirled around to confront whoever it was, too angry now to consider the consequences.

It was Moipa, Dabo's youngest son, and our workmate. He froze the moment I turned. "Simon?" he asked, before mumbling something in Huao and tapping his leg. He pointed to our camp.

"Come on," I said, gesturing him to follow me.

I'd grown close to Dabo and his boys, and although we had no real language in common, we were able to communicate pretty well by this stage. Which was just as well considering Tyaento spent less than an hour every few days at the dredging site.

I showed Moipa to Simon's room. Everyone welcomed him in, except Simon, who, in his fever, looked through us all. Moipa nodded, his expression grave. He poked Simon's legs with his finger, probing around the wounds and causing more pus to ooze out. Simon grimaced and breathed rapidly through his nose. I tried not to breathe at all. Moipa smiled at Simon, nodded again to the rest of us and then left.

Henry and I watched over Simon while Iveta took a break. When she came back in, we bid her goodnight and went to our tents. As

Falling to Fly

worried as I was for my friend, I was exhausted after the day's dredging and fell asleep easily.

Sometime later, I drifted awake to the sound of shuffling feet and an urgent knocking. Assuming it was a drunken Huaorani stealing food for a midnight feast, I lay there fuming instead of moving.

"Don Simon?" a voice whispered outside. It was Dabo.

I switched on my flashlight to check the time. It was four-thirty in the morning; too early for work. What was Dabo doing here? And why was he trying to wake Simon and not me or Henry? He knew Simon was injured.

"Go away!" Iveta said. Her voice muffled behind the door.

"Tengo ayuda," said Dabo. *I have help.*

I poked my head out the tent. Dabo was with Moipa and Kimo, and a tiny old woman bundled up in rags. I'd not seen the woman before. I would have remembered if I had. Her eyes were piercing and coal black and her mouth was a puckered toothless depression in the lined topography of her face.

"I don't understand you," said Iveta. "It's late. Go away."

I pulled on my clothes and crawled out to greet Dabo and his entourage. "Que pasa?" I asked. *What's up?*

"Llevemos la curandera para Don Simon," he said with a broad smile. He pointed to the little woman who dipped her head and shoulders in a polite bow.

"Iveta, it's Jason. Open up. These guys are cool. It's Tyaento's brother-in-law and nephews. They've brought an old woman who they say is a healer."

Iveta cracked the door a couple of inches to peer out. Dabo pushed it open and we all poured into the room.

"How is he?" I asked Iveta, over the chatter of the four Huaorani crowding around the bed.

"Same. Not good."

"Se necesita fuego," said Dabo. "Para preparar el té."

"What's he saying?" asked Iveta. To her credit she was very calm, all things considered.

"He needs a fire so Yoda's mom can make some tea," said Henry from the doorway. He was still in his underwear. "I imagine a camp stove will do though."

"Tea? What the hell, guys! Do they seriously think I'm going to make tea?" Iveta was standing with her hands on her hips now, her expression incredulous and defiant - so much for being calm.

"No, they don't," said Henry. "The old crone wants to make the tea herself. An herbal brew of some kind that'll help Simon."

"Oh." Iveta smiled at the old lady and lit the camp stove on the table. "Gracias."

The woman's eyes went wide, and she nodded approvingly at the efficiency of modern magic. She then muttered something to Kimo and took a paper bag out of a fold in her clothing. Kimo filled a large pot with water and put it on the stove. The old woman emptied the contents of the bag into the pot. I peered in. There was variety of leaves and splinters of tree bark, some spongey and mustard-colored, others black, hard and shiny like obsidian.

While the potion was heating, the woman returned to Simon. She stripped back the sheet, gently removed the bandage covering his leg, and began cleaning his wounds with a damp cloth. From nowhere, a small plastic soda bottle filled with a dense burgundy liquid appeared in her hand. She emptied a glob of the syrupy substance into her palm and began rubbing it with two stubby fingers. The blood-red ooze transmuted into a pink foam, which she then applied to the open sores. Repeating the process several times, she made sure to cover all the lacerations and surrounding skin. Once she'd finished, Simon's leg looked as if a rabid dog had been chewing on it, the pink froth drying to a crusty dark red. Iveta looked perplexed, but Dabo's reassuring smile was enough to convince us the old woman knew what she was doing.

She made Simon drink a cup of the bitter smelling brew she'd prepared, and then pressed the cup into Iveta's hand; passing back the responsibility of Simon's care to his girlfriend. More paper bags filled with the tea mix materialized from the folds of her enchanted rags, and she placed them on the table by the stove. She positioned the soda bottle at the end of the row of paper bags and then turned to Dabo.

The woman was either giving him rapid instructions or subjecting him to an intense interrogation. Dabo nodded, face grave, and gave an occasional monosyllabic answer to her barrage of words. When she finished, the two of them embraced and Dabo gave the old woman a few flakes of gold, three packets of instant soup and a candy bar. She smiled her gummy grin and secreted her payment away within the pleats of her tatty skirt. After hugging us all in turn, she left without another word or a backward glance.

"That was entertaining," said Henry. He gave Iveta a look of sympathy. "Let's hope it works."

I picked up the soda bottle and took a closer look at the viscous red goo. "What is this stuff?"

Henry turned to Dabo. "Que es esto?" he asked, pointing to the bottle in my hand.

"Sangre de Drago."

Henry chuckled.

"What did he say?"

"He said it's dragon's blood. Which would suggest the potion probably contains eye of newt and wing of bat."

*

With some difficulty, Dabo relayed the old lady's instructions to Henry, who in turn passed them on to Iveta. Simon had to drink the tea, and nothing else, until his fever passed. The dragon's blood had to be reapplied as often as necessary, especially after washing out the wounds. If he didn't get better in few days, Dabo would fetch the woman back.

"Where does she live?" I asked.

"Five hours away," said Henry. "He didn't specify if that was by canoe or foot. Either way, they must have left as soon as Moipa got back from visiting us yesterday evening. They traveled all night to bring her here. I told them to go rest. Said we can handle the dredging today."

"What about taking Simon to the hospital?"

Henry glanced at Iveta, who was bent over Simon, mopping his brow with a wet rag. "Well, Tyaento won't let us have a canoe so getting him out is going to be tough. Dabo swears by the old woman and her potions – there's no doubt in his mind Simon will get better now. I say we wait until this afternoon. If he's still bad, I'll take him out, even if I have to steal a canoe to do it."

Iveta gave Henry a tight little smile. "See you this afternoon. Please don't be too late. We'll need daylight to carry him in the jungle. I don't want to steal anything from these people. They scare me."

*

It was too late to go back to sleep so we walked to the river and waited for the sun to rise. The sky paled and the night sounds faded to a few scattered calls. For the briefest moment there was utter silence, as if the forest was taking a deep breath, bidding farewell to the night and welcoming the new day. Henry and I were silent, too. The morning was immaculate and achingly beautiful; every plant and tree vivid and new. The morning birdsong, when it came, was sparse, muted, respectful.

Kimo arrived and the three of us began work, shattering the peace. The dredge's motor clattered and the air compressor rattled and hissed as we sucked holes in the riverbed. I dived with a plastic pipe in my mouth that fed gasoline-flavored oxygen to my lungs. The hose bucked in my hands, churning up sand and silt. I couldn't see a thing, but at least it was quieter beneath the surface of the water.

We were working a very rich vein and by midday we'd made more than half an ounce of gold. It was the first time we'd stood a chance of reaching a whole ounce in a day, but we had to finish early for Simon. I wasn't sorry to stop. My head was pounding and the delight of the still morning was now a distant memory. Apart from seeing gold mount up fast, there's no real joy to be had in dredging.

When we returned to camp, Simon and Iveta's door was open and I could smell something good cooking. Iveta came out to meet us, beaming and excited. The difference in her countenance from the dour and anxious woman we'd left early in the morning couldn't have been more marked. The change in Simon was even more striking.

He was sat up in bed, eating soup. His skin was still pale and taut, but he no longer had the waxen hue of a corpse. Best of all, the putrid smell of rotting skin was gone. His leg still looked terrible, but the dark marbling had receded, and it was now merely warm, not scalding, to the touch; the furnace of infection had been doused.

"Iveta tells me the toothless goblin wasn't a hallucination, and I've got her to thank for fixing me up," he said, taking a sip of tea and pulling a nasty face. He then raised his cup and grinned. "To our good health gentlemen!"

*

Three days later Simon was working in the garden. Within five days his leg was healed and he was back working the dredge. All that remained of the wounds we'd believed would cost him his leg, if not more, was some faint scarring.

The healer's bottle of dragon's blood only lasted a couple of days, but Dabo was able to acquire more with ease. He took me with him into the jungle where he found a tree with heart shaped leaves and crisscross scars up and down the trunk. Taking out his machete, he cut into the bark and let the tree's scarlet sap run down the blade into a jar.

"Sangre de Drago," he explained, holding up the jar once it was filled.

By now, I was growing pretty smug about my constitution. Beyond some mild diarrhea, I'd stayed pretty healthy all the time we were in the jungle, while Henry and Simon, both bigger and stronger than me, had been laid low for days at a time. Iveta also avoided any kind of sickness or injury, but she'd stayed in the village while the rest of us worked long, hard days on the dredge, often imbibing dirty river water mouthfuls at a time. In the end, it turned out I had no more immunity to the harsh conditions and foul microorganisms than the other guys, and when sickness did eventually hit, it knocked me out cold.

I'd been feeling progressively worse all day with nausea, headache, and a foggy brain. Like an idiot, I'd kept working. It wasn't until I fainted on the riverbank that I was able to admit something serious might be wrong and began to stagger and stumble back toward our camp.

Simon came after me and helped me get there. When I arrived, my vision was blurred and colors whirled in the corners of my eyes. I vomited on the steps up to the porch. Simon said it didn't matter and kept me walking to my tent. It was the middle of the afternoon and the sun was scorching overhead. I unzipped the door flap and was hit by a blast of hot, stale air, stinking of plastic. I threw up again and then crawled into the relative cool of the storeroom, where I curled up and shivered and sweated.

People came and went, swimming in and out of my field of vision. Some were only feet, dirty and gnarly. Others I recognized; Simon, Iveta, Henry. I was dimly aware of my friends trying to take care of me. Jugs of water and cups of tea appeared on the floor beside me. Later, soup. A feather-light alpaca blanket was draped over me when it got dark. I couldn't keep anything down, not even the healer's tea, and at some point in the night I soiled my pants. I didn't sleep. Instead,

I drifted between brief and intense fever dreams and long, dull, painful wakefulness.

In the too-bright morning, Simon and Henry washed me and changed my clothes. Iveta brought more tea and changed out my sick bowl. I lay trembling, wanting to die. More people came in and out of the storeroom, laughing and nudging me with their feet as they stole our food. Day became night, then day, then night; time was liquid, irrelevant.

I floated and shook, so cold, then too hot. Creatures sniffed and growled at me, and I knew they were waiting to gnaw the flesh from my skinny bones. They were welcome to it. If only I wasn't so afraid. Sharp pain stabbed into my right kidney. I gasped and wheezed, my mind jerking back from a nightmare of vultures and rats, fighting for my soft gelatinous eyes. I was back in the storeroom. All was bright and blurred.

"Lazy little termite! What are doing in here?" Tyaento yelled, kicking me in the back again. "You want to stop the people from eating by making yourself the guard dog?"

I twisted my head to look at him. I've never hated anyone more. I opened my mouth, but nothing came out. My throat was dry. My eyes were damp, filled with tears of pain and frustration and loathing. Tyaento took a big puff on his joint and blew smoke in my face.

"Are you crying, gringo pussy? You people are so weak. We should spear you all like the old days. Ha!" He jabbed my chest with stone-hard fingers. I flinched and grit my teeth. "But no, we're smarter now. We learned a lot from that bitch Rachel Saint and her churchy fuckers. Or at least I did. Now we have everything because you stupid bastards give it to us. You come for oil and gold? You can have it, but Tyaento makes sure we get paid. Tyaento knows how to play the game from all sides. I'm too clever for you termites. I'm the termite who eats the termites!"

I couldn't listen to him anymore. I rolled over and put my arms up over my ears. He kicked me one last time and then his footsteps retreated. I lay on the dirty wooden floor and wept. I wasn't sure why - maybe for the pain in my kidneys, or in my head and neck and stomach, or maybe it was the fever, or, most likely, the sure knowledge that we were being used and abused. We'd come to help these people, and all they did was take from us and laugh about it. We'd been played from the start.

"Jason, I'm going to take you out of here," said Henry. Simon and Iveta were behind him, one behind each shoulder like benevolent angels. "The tea's not working. Neither are the antibiotics. We don't know what to do. Dabo went to find the witchdoctor, but she wasn't home. You're going to waste away if you stay here."

I nodded my assent and Henry and Simon lifted me to my feet. My head spun, but I managed not to vomit on my friends.

"Can you walk?" asked Henry.

I nodded again and felt my brain slosh and smack up against the inside of my skull. I wanted to scream but didn't have the energy. Outside the storeroom, the wooden platform was bare. Our tents were gone. I couldn't have cared less. I leaned against Henry and walked, slow and steady, out and away from the village.

The trek back through the jungle was one of the hardest things I've ever done. It was a grueling slog without end. But I didn't want to stop; I wanted to be as far away from Tyaento and his corrupted Huaorani as possible. If I never saw them again it would be too soon.

It wasn't until we reached the dirt road at the end of the track that I realized Dabo, Moipa and Kimo had walked with us, carrying our backpacks and tent-bags. Looking at their familiar, concerned faces made me feel guilty about the thoughts I'd been having about the Huaorani. These guys were okay. In fact, they were more than okay; they were our friends.

A white pick-up truck taxi bumped up the track. I hugged my Huaorani friends, who then helped me up onto the truck's backseat, where I collapsed and curled up into a ball. Henry climbed in the front, and we left the jungle for the hospital in Puyo.

The doctors didn't know what was wrong with me. "Maybe dengue, maybe amoebic dysentery, maybe both," said one doctor, before prescribing an intravenous drip of some kind, and super-strength antibiotics. After a couple of days I felt well enough to leave and go to a hostel, where Henry continued to take care of me. Simon and Iveta had stayed on in the Huaorani village alone.

Just when I thought I was getting better, I had a relapse. The fever and diarrhea returned and I was bedridden again. It was while in this weakened state that I received a visit from Simon and Iveta, fresh from the jungle.

"We need to do something about that bastard," fumed Simon. "He got even worse after you two left. It was unbearable, especially for

Iveta. First, they took our food *after* we'd cooked it. Then our share of the gold we'd made since you left went missing. I confronted him about it and he told us to leave. They threw us out of the village! I asked about the dredge and he told me not to touch it. Can you believe it? I'm sorry, Jason, I know how much it cost you, but he just refused. Said if it was in Huaorani land it was their property. I would have strangled him if he hadn't been surrounded by the whole damn village!"

"It's not the cost," I whispered. "It's him. I'd rather burn it than let him have it."

"Really?" said Henry from the back of the room.

I nodded, which set off a violent coughing fit.

Once I'd finished, Henry said, "I feel the same way. But I'm not sure we should burn it. It'd be a shame to waste all that hard work. Not to mention the fact it's ours."

"What do you have in mind, Henry?" asked Iveta. "If we can screw over the bastard pervert I say we do it."

All eyes were on Henry, who grinned. "I think we should steal it back," he suggested. Then he glanced at me. "When you're feeling better, of course."

"We can't wait that long," said Simon. "He might move it somewhere else by then. You know how devious he is. We need to go now."

"Tomorrow night," I suggested. "We need to arrange things properly, and we'll need a day to do it."

"There's no way you're coming with us," said Henry.

"Agreed," said Simon. "Not with that cough. You'll wake up the whole village."

"Let's see tomorrow night," I said. "In the meantime, one of you needs to change up some of my gold for cash and go to Tena."

Henry went to Tena and hired a truck and six guys who were willing to work all through the following night. We couldn't risk hiring people in Puyo, as there were too many people there who knew Tyaento. Simon and Iveta went to Misahualli, a small touristy town north of Tena, and spoke to a guy I knew who rented storage space. By the following morning, they were all back and everything was set.

It was clear that I was still too sick to go with them, so I stayed behind with Iveta, while Henry, Simon and the hired hands set off in the truck. They left at six in the evening. Iveta and I stayed in our rooms and fretted all night. Drifting in fever, I kept dreaming up the most terrible scenarios. By morning, I was convinced Tyaento had already caught, tortured and killed my friends, and was now on his way to take his revenge on Iveta and me.

At ten o'clock, the handle to my door rattled and I jumped off my bed and ran to the window. Fever or not, I had no intention of being speared by Tyaento. I wrestled it open and was about to climb out when I heard Simon's voice.

"Open up, Jason. It's time to go."

I staggered back across the room and cracked open the door. Simon and Iveta were standing there with their backpacks. Simon was smiling. Iveta looked relieved.

"We've got it," said Simon. "Now let's get the hell out of dodge."

The four of us rode to Misahualli in the back of the truck with the dredge. Henry and Simon were exhausted but elated. Simon in particular was a ball of nervous energy.

"It was incredible," he said. "We got to the river after dark and waded all the way down, past the village, to where the dredge was tied up. It hadn't been moved at all. I don't think they'd even used it since we left. I'm sure the arrogant asshole didn't suspect for one minute we'd have the balls to do what we did. We pushed and scraped that six-hundred-pound sucker all the way up the near dry river. Shrimp were hopping and plopping all around, snakes were slipping up against our legs, and I was sure someone was going to hear us. But no one did. We got it out. I can't believe we actually stole it back from the bastard!"

"Me neither," I said. "You guys are amazing. I owe you big time."

"You don't owe us anything," said Henry. "We came for an adventure and you delivered it. My only regret is that none of us was there to see Tyaento's face when he realized the great Huaorani prince was outwitted by a handful of scuttling gringo termites."

Jason Pednault   Matheu DeSilva

Tequila

We spent a nervous couple of weeks in Misahualli. Holed up in a rundown hostel on the wide sandy river-beach that doubles as a ramshackle port for motor-canoes taking people and goods to the remote villages and luxurious lodges further in the jungle.

We chose Misahualli because it's a tourist village, slightly off the beaten track, populated predominantly by indigenous Quichua, with a smattering of Europeans and mestizos. Traditionally, the Quichua keep their distance from the Huaorani, and vice-versa, so we figured it would be a good place to hide out with our fugitive dredge.

My friends enjoyed the village restaurants and the semi-domesticated monkeys terrorizing tourists in the square, they hung out on the beach, and took a few trips downriver - always with an eye out for Tyaento and the Huaorani - while I stayed beneath the mosquito net in my clapboard oven of a room, or outside on a hammock strung between shade trees. The sickness I'd picked up was not going away no matter how much medicine I took. I was the weakest I'd ever been.

In the evenings, Henry and I discussed the places I'd prospected in Ecuador, trying to decide where best to take the dredge next. Our plans were half-hearted though. I was in no fit state to go anywhere, and Henry's visa was about to expire.

Two weeks after arriving in Misahualli, we all went our separate ways. Henry went back to the States. Simon and Iveta travelled south to Peru. I went back to Vilcabamba.

*

"You're back!" said Kitty when she spied me sitting outside the juice bar.

"Surely you knew I was on my way," I said, taking a sip of green juice.

"No, I didn't." Her face took on the detached glaze I'd gotten used to as the prelude to a psychic revelation. "In fact, I'd say you're not supposed to be here for a few months at least."

"Well sorry to disappoint the oracle, but I needed to get back. This is still the center of the healthy living and healing world, right? I could do with some of that about now."

"Right," she said, looking me up and down. This time she appeared to actually see me and not my aura, or whatever she usually saw. "You do look like shit on toast. Do you want me to do some reiki for you?"

I laughed, having no idea what reiki was. "Why not. Will it help with tropical fever?"

"Of course it will. It's energy healing. It helps with everything. What are your symptoms?"

I told her about the recurring fevers, stomach cramps, bouts of vomiting and diarrhea, the coughing fits. I also explained that the antibiotics I'd taken only seemed to work for a few days before the symptoms kicked back in again with a vengeance.

"Are you taking antibiotics now?" she asked.

"Nope, just green juice, as prescribed by nearly everyone I've seen this morning."

"Good. On both counts. Antibiotics should only be considered in extreme emergencies and not popped like skittles, as recommended by most doctors. I'm no expert, but it sounds like you have amoebic dysentery. Have you tried taking dragon's blood?"

"Dragon's blood? How do you know about that? I only discovered it a few weeks ago."

She raised an eyebrow. "You discovered it did you? Next thing you'll tell me Columbus discovered the Americas."

"You know what I mean. And he did, didn't he?"

"Many people discovered it before that idiot set it on its present course of destruction and delusion. Not least of all the first settlers, or Native Americans as they're now quaintly called."

"Kitty, I'm not really in the mood for a lecture on the noble indigenous of the Americas, north or south. They're the reason I'm in this mess. Well, one of them is anyway. So, dragon's blood?"

"Sorry, yes, dragon's blood. It's been used to treat everything from cancer to hemorrhoids. Ask Ravi. He used it when healing his so-called 'incurable' illness a few years back. So you know we're talking real healing chops and not just the bum-grapes end of the spectrum. It's quite the panacea, and one thing it's particularly good with is gastrointestinal problems." She tapped my tummy. "Like yours."

I flinched. "Do you have any?"

Kitty shook her head. "Neither does Ravi. He ran out, too. Someone was looking for some a while ago, but no one in town had it. You can get it in the jungle easy enough though. If you know what the tree looks like and have a machete you can get it for free, too."

"Thanks Kitty. Looks like I'll be heading back to the jungle tomorrow, then. If I don't see Ravi and Dana before, send them my regards. I'll see you in a couple of weeks."

"Months," she said, correcting me. "Take care, Jason. Make sure you build up your strength. You never know when you might need it."

*

It took three separate buses to get to Tena. On the second, the fever descended again, making it a minor miracle I caught the third. I arrived at the Tena bus terminal in a state of burning disorientation. Staggering and forgetful - I left my backpack unclaimed in the storage beneath the bus - I stumbled toward the marketplace, looking for dragon's blood.

"Señor Jason," said a voice I recognized. Though from where, I wasn't sure.

I turned to see who it was and the world swirled, making me topple over and crack my elbow on the sidewalk. The pain was sharp and made bile rise in the back of my throat. The air all about was hot and clammy, but the concrete sidewalk was cool and welcoming. I laid my cheek against it and closed my eyes.

A jolting movement woke me, irregular, up and down, shaky. I was being carried through the streets; a man held my arms and a short, stocky woman grappled with my legs. I didn't care what was happening, or where they were taking me, I just wanted to sleep.

The second time I awoke, I was in a familiar room. The white walls, the noisy air conditioner, the sweeping overhead fan and tiger print bedspread; I recognized it all. Somehow, I'd made it to the small hostel where I stayed whenever I came to Tena. I tried to think how I could have gotten there. The voice; it had been Pablo, the hostel's owner. He must have found me wandering the streets and brought me here when I fell.

I climbed out of bed and looked around for my things. There was nothing apart from my vomit-stained shirt on the floor of the shower, and my shoes, neatly placed at the end of the bed. Dizzy, shirtless, barefoot, filthy and emaciated, I made my way down to the reception desk, torn between the hope nobody would be there to see me and someone being there who could help locate my missing stuff.

Luck was on my side; no one was at the desk, and behind it, leaning against the wall, was my backpack. I guessed Pablo must have retrieved it from the bus station. The fact it had still been there was a true wonder in South America. I took hold of the straps and dragged it bumping up the stairs to my room, where I collapsed in a sweaty heap and went back to sleep again.

Pablo's wife, Carmen, came in to check on me the following morning. She brought me eggs and bread and a pitcher of water. She also had a small bottle of dragon's blood.

"Cinco gotas en el agua," she said, pouring a glass of water and adding five drops of the heavy red liquid. "En los mañanas y la noche."

"Gracias," I croaked. Carmen smiled and left the room.

I understood she'd told me to take the dragon's blood in water twice a day, morning and night. But what I couldn't understand was how she'd known I'd come to Tena for dragon's blood. Maybe Kitty had sent a telepathic message. I grinned at the thought, and downed the sap-infused water. It was bitter and faintly rubbery, and left my mouth feeling sucked of moisture, as if I'd bitten into an unripe persimmon.

Within a few days my stomach stopped cramping and the fever was gone. I still felt weak but was no longer sick. After two weeks, the symptoms hadn't returned and I was getting my strength back. The sap of an obscure jungle tree had worked yet again where modern medicine had failed.

During my convalescence, I spent time getting to know Pablo and Carmen and their daughter, Belén. My Spanish was nowhere near as good as it should have been considering how long I'd been in Ecuador, but we managed to communicate well enough.

They told me about the many herbs and elixirs from the rainforest used in traditional Ecuadorian medicines. Carmen's grandmother had been a healer, and Carmen had worked with her before marrying Pablo. That's why she'd known I'd needed the dragon's blood.

She spoke about passing her knowledge on to her daughter, but Belén, being a teenager, was of a more modern mindset. She shunned old-fashioned herbal remedies in favor of whatever was being marketed at the pharmacy, whether it worked or not. Carmen hoped once Belén matured she'd come back to the old ways. I hoped so too.

Once I was back to full health, my mind turned to what I would do next. I didn't have much money left, but I did have most of my prospecting gear with me. I also had the dredge - although the idea of using it left me cold; it was noisy, polluting and obnoxious. The mere thought of it took me right back to Tyaento and the sickness I'd only just gotten over.

Even though it had cost me nearly seven-thousand dollars, I decided to let the dredge go for free. Trying to sell it would bring too much unwanted attention. Besides, the man who was storing it for me was a widower struggling to raise several children on the pitiful income he made off the dilapidated storage sheds behind his shack. I imagined owning a dredge would be life changing for him and his children, so I went to Misahualli and told him he could keep it.

With the dredge gone, I felt lighter and ready for another adventure. Being back in Tena helped me realize where it was I wanted to go next. It seemed Kitty had been wrong about one thing: Vilcabamba wasn't the only place in Ecuador I felt at home. There was also the Jatanyacu.

I went into town to buy a pry-bar and some other heavy-duty tools for cracking open rocks (on Henry's advice). When I arrived back at the hostel, Belén was waiting for me outside with a tiny puppy cradled in her arms.

"Para usted," she said, handing me the scruffy white bundle of bones.

It looked like a miniature mop head with a teddy bear's face. "Gracias," I said, holding him up for closer inspection. He was one of the cutest - and smallest - puppies I'd ever seen.

A few days earlier I'd told the family about the dogs I'd had growing up, and how invaluable they'd been in the wilderness. I was nervous about returning to the Jatanyacu after my ill-fated trip with Ravi and wanted to get a dog to help set my mind at ease. Dogs have a sixth sense for danger, and in the jungle in the middle of the night an early warning of flood or a predator could mean the difference between life and death. Pablo had said they'd help find me a puppy I could train. Now they had, and I was grateful, even if he was a good deal smaller than what I'd envisioned.

Belén held her hand up to the pup and giggled while he licked her finger. "Su nombre es Tequila," she said.

"Tequila," I repeated. How on earth he'd come by that name I couldn't imagine, but this was Ecuador and names were often pretty odd. I was just thankful he wasn't named Hitler like the mutt down the street.

"Si." Belén nodded. "Es Tequila."

The dog's tiny ears pricked up and he wagged his tail at the sound of his name. Tequila it was then. We went inside, and Belén helped me fashion a dog-hammock out of one of my old t-shirts.

*

"So, you trying to pick up chicks with that thing?" said a young woman with long mousy hair and a soft, melodic accent. Without waiting for a response, she plonked herself down in the chair opposite.

I glanced around the cheap local restaurant where Tequila and I had been eating lunch for the past couple of weeks. It was full. Lunch hour for the municipal offices across the street must have just begun. People were waiting in couples and small groups for tables to become free. I guessed this woman hadn't wanted to wait.

She ordered her food in perfect Spanish and then turned her attention back to me. "So, are you a mute, or shy, or stupid?"

"Do I have to pick one?"

"Pick as many as you like. Why you carrying around a puppy? Are you a pedophile?"

"What?"

She chuckled. "Chill out! I'm only pulling your leg. I'm Amy, by the way. I work at the rafting company across the street. You interested in going rafting? Or do you just like playing with little doggies?" She had such a mischievous smile it didn't matter what she said, it was impossible to be offended.

"I've never really thought about rafting," I admitted. "I like going out fishing. My dad has a boat in Alaska."

"Bully for him. So was that a yes or a no?"

"To what?"

"Rafting, you dullard."

"Not at the moment."

She threw her head back and cackled like a witch. The people on the next table shot her a nervous glance. "Of course not at the moment. But while you're here in Tena, I mean."

I shrugged. "Maybe. Like I said, I hadn't really thought about it."

"Then why are you here? Most people come for the rafting, or the jungle tours. And they don't often bring their pets."

"I'm a gold prospector."

She gave me a skeptical look. "With a puppy?"

"Yep."

"So you've come to pollute the rivers, huh? Great. Well, at least you've got a puppy so people will *think* you're a nice guy, when in truth, you're really a prize asshole."

"What? I'm not here to pollute the water."

"You think the mercury's good for it? If I'd known you were a gold man I wouldn't have sat here."

"I didn't ask you to sit at my table. And I don't use mercury. I love nature."

"But you like taking gold out of rivers?"

"It's how I make my living."

She glared at me for a full ten seconds, then shrugged and said, "I guess we've all got to do something."

"We do," I agreed. "Speaking of which, how long have you been a rafting guide?" I was trying to divert attention from me and my apparently evil profession.

"Years," she said. "I was born into it."

"You must have been. You look too young to have been doing it for years."

"Oh, you're a charming bastard now, are you? Not only do you have a puppy, I see you're also having a stab at the gift of the gab. I know I only look twelve, but I'm actually twenty-four."

"I didn't say you looked twelve, just that you didn't look old enough to have been rafting for years."

"Well I have. My family's been here since I was little. My parents had the bright idea to come here from Ireland and preach the word of the Lord to the natives. I grew up in Tena, kayaking and rafting these rivers with my dad and brothers. So if you ever do want someone to take you out there, give me a shout. We're the best in town, even if I do say so myself."

"Do you ever go out to the Llanganates Park? To raft the Jatanyacu?" I asked.

"Ah, so you know the rivers here already. As a matter of fact we go there all the time. Why do you ask?"

"Maybe you can give the two of us a ride next time you go out there." I pushed Tequila up in his make-shift hammock so he could turn his adorable little gaze on Amy, and then added my best puppy-dog eyes to the tableau for good measure. "The taxis there and back are pretty steep."

"Ha! Will you look at the two of you! At last, the truth of why you carry the little shite about: you're a double act! And all because you're too tight to pay taxi fares. Who'd have thought it? Of course I'll take you out there, and for nothing, too. As long as you promise not to pollute any of my rivers like all the other gold prospecting dick-heads."

"I promise." We shook hands, and I paid the two dollars for her lunch to seal the deal. The way my money was dwindling, free rides in and out of the Llanganates were a godsend. Seemed like Tequila was turning out to be my lucky charm.

*

Three days later, I hauled all my kit down to Amy's office at six in the morning. Tequila followed along at my heels. He was turning out to be a great little dog. Smart and obedient, and potty-trained already. I'd taken him on a few hikes outside of town since I'd gotten him, but this was going to be the big test for his little legs. I hoped he wasn't going to find the jungle too tough.

There were three couples from Colorado already waiting in the office when I arrived. They all made a fuss of Tequila, cooing over his dinky size and asking to pet him. Amy rolled her eyes and winked at me.

At six-fifteen, a large panel van with a raft and a kayak strapped to its roof pulled up outside. The rafters climbed in the side door while Amy and Juanito, the van's driver, helped me load my stuff into the back next to the piles of lifejackets and paddles.

"Who's the kayak for?" I asked when we were on our way.

"Me," said Amy. "There was a storm up in the mountains last night, so the river should be nice and gnarly today. Juanito can sit in the raft with this lot while I have some fun."

"You're not coming out on the river with us?" one of the guys asked me.

"Nope. Indiana Bones here is going treasure hunting," said Amy. "Aren't you, my skinny, dog loving friend?"

I glared at her.

"No shit? I wondered what you were all loading in the back."

"I'm just doing some prospecting," I said. "It's no big deal."

The last thing I wanted to do was explain my crazy lifestyle to a bunch of fellow Americans, fresh off the plane. Thanks to Amy, though, that's exactly what I did for the next half hour. They were fascinated and kept asking me questions. How? Why? And wasn't I afraid to camp out in the jungle alone? While I mumbled answers and explanations, Amy sat beside me, chuckling and smirking.

Juanito drove right up to the park entrance for me. He and Amy hopped out to help me unload my kit and supplies.

"How long are you planning to stay out here?" Amy asked once we'd got everything out.

"Four or five days. Then I'll go back to Tena to stock up on supplies again."

"Which is it, four or five? If you're wanting a lift back to town you'll have to be more specific."

"You don't have to-"

"I know I don't, but I'm offering. No skin off my nose to come out here. This is one my favorite rivers anyhow. You want a ride back to town or not?"

"Sure. Thanks Amy. I appreciate it."

She rolled her eyes. "I don't have all day."

"Right, sorry. Let's call it four this time."

"This time? You're a hopeful get, aren't you?" She grinned and jabbed my shoulder. "Four days it is. Be here this time Thursday morning and we'll pick you up. Good luck."

"You too," I said.

"Ha! Keep it for yourself. I won't need it."

She climbed back into the van and slammed the door shut. After they sped off I noticed nobody else was around. The wooden hut where they sold the fried fish and yucca was closed up and there was nobody in the pools. Perfect timing.

I had the crumpled permission slip we'd bought from the Indigenous Cooperative in my pocket, but still I'd hoped by coming out so early on a Monday morning I'd be able to sneak into the park unobserved. I didn't want a repeat of the attention we'd received from the various 'owners' of the land last time I was here.

I loaded myself up and walked across the swinging bridge and into the jungle. Once I'd made it a fair distance in, I looked for a place to stash my big backpack and water jugs. I planned to continue on with only my daypack and find the best spot to camp and pan, as far from prying eyes as possible. Then come back for the rest of my gear.

I was pleased to see Tequila taking the jungle in his tiny stride. He was darting back and forth, sniffing and wagging his tail. He had grown to the size of a full mop head now and was happily soaking up mud like one. I dragged my pack and the string of water jugs under a bush and began hacking my way further into the undergrowth with my machete. Tequila cocked his head, then barked a warning and backed away growling. I glanced around but couldn't see anything to worry about.

Then I heard the low thrumming of a swarm of bees or wasps. The sound grew louder and I crouched, searching in the dim light beneath the canopy, trying to spot them. A series of burning pinpricks rippled up my forearm and I dropped my machete like a hot rock. The creatures were tiny and black like gnats, but they moved and buzzed and stung like wasps. I hustled out from under the bush and found Tequila on the trail, head cocked, gazing up at me.

"Yes, I'm an idiot, and I should have listened to you."

Tequila barked in agreement and shook his rump.

Lesson learned, I slathered the raised welts on my arm in dragon's blood and found a different place to hide our kit. With my load lightened, I was able to fully sink into the sights and smells of the jungle. It was so good to be back. Tequila scurried beside me, over rocks and fallen trees, showing no signs of tiring.

We reached a fast-flowing creek and I stopped to wash my face and neck while Tequila took a drink. The creek wasn't too wide and I let Tequila figure out how to hop across the stepping-stones behind me. He almost fell in at one point but was able to scrabble up to safety. When he reached the shore, he did an odd little dance around my legs, prancing like a deer while wagging his ratty stump of a tail.

We took an overgrown path down to the secluded beach and then the two of us headed upstream over boulders and slabs of rocks. Tequila found a hidden clearing just above the high waterline on one of his forays into the bushes. It was a perfect campsite.

"We're quite the team, buddy," I said, and gave him a piece of bacon from the plastic bag I kept in my pocket.

Once I'd cleared the area and gathered a few twigs of firewood, I set off to retrieve the rest of our supplies. Tequila followed, but reluctantly. Whereas he'd scampered ahead on our way in, he now hung back, unwilling to go any further than he had to. I thought he might be getting tired, but once I loaded up our stuff and started back to camp, he brightened up and bounded like before.

"Did you think we were leaving, boy?" I said. He barked once in answer. "I guess you must love it here as much as I do, huh?"

It didn't take long to set up camp. I strung up my hammock, with its built-in mosquito net and rain cover. Then fixed a tarp above it, just in case. No tent necessary. I stashed the firewood under my hammock and unpacked my prospecting gear, ready for the morning. Tequila and I shared some bread and a couple cans of tuna for dinner, and then climbed into the hammock and went to sleep.

He whimpered in the night, and I awoke, straining to hear the sounds of a storm or some other danger. It turned out there wasn't any. Just a need for a little dog to empty his bladder. Once he'd finished and was back in the hammock, I lay there listening to the frogs and the crickets and the river until I drifted back to sleep.

In the morning, there was something stuck to the side of the mosquito net. Blurry eyed, I thought it might be a bat, or maybe a large leaf that had fallen in the night. I eased myself away from it, and rummaged for my glasses. As soon as I put them on, the object opened up, revealing wings of iridescent cobalt. The giant blue morpho butterfly lurched upward and flew in its clumsy, lopsided fashion twice around our camp before heading away toward the river.

I started work early, following Henry's advice to split the rocks into layers to get at the deep deposits. Tequila chased butterflies and birds while I panned. It was idyllic; calm and peaceful and productive. Wednesday night arrived all too soon. I didn't want to go back to Tena, but I imagined Amy would worry if I didn't make our rendezvous in the morning. Besides, I needed to cash in the gold and buy some more food. For a small dog, Tequila definitely got through more than his fair share.

Thursday morning, I broke camp and hid everything away, then hiked out to meet Amy. She and Juanito were there with a van full of rafters. We dropped everyone off a little way downstream and then Amy drove Tequila and me back to Tena.

The next few weeks followed the same pattern. Amy, or one of her brothers, would drop me off at the Jatanyacu. Either they'd come back and pick me up a few days later, or I'd catch a ride in with someone who had visited the pools at the park entrance. I'd cash in my gold, and usually stay in Tena for at least a day or two, hanging out with Amy and her rafting friends, or other tourists I'd meet around town. Tequila, of course, would come everywhere with me, lapping up life as the center of attention.

Jason Pednault   Matheu DeSilva

Peligro

It often rained at the Jatanyacu, especially in the evenings. If the rain continued through the night, the mornings would be misty and magical. The river would rise and flow, bringing more gold to the cracks and crevices of my hunting grounds.

Rarely would it storm, and never so hard as to bother me beneath the tarp and flysheet in my hammock. It was a different world to the crashing, lashing night of panic I'd experienced with Ravi, Gabriel and the others. Every time the heavens opened, however, I always experienced a slight pang of fear. I knew what the Jatanyacu was capable of and didn't want to be caught out again. I made sure to always set our camp up far above the high-water line.

One Friday afternoon the sky to the west darkened, obscuring the distant mountains in a vast smudge of purple-gray clouds. I'd only been out for a couple of days but had already made the quota of gold I'd set for this particular trip. I could afford to stop early and snuggle up and watch the storm roll in. Tequila read my mind, and by the time I'd stashed everything away he was already sitting beneath the hammock, barking to be lifted inside.

We ate an early supper in the hammock, cocooned against the pitter-patter of rain. The sun still shone on the river; somewhere out there would be a rainbow, stretching over the canopy that shaded our camp.

The early dusk of the storm rolled in, bringing a heavy curtain of rain. The river began picking up speed and rising. Safely ensconced in my hammock, I listened to the fat drops drumming on the leaves and on my tarp. By the time real dusk arrived, Tequila was sound asleep and I wasn't far behind.

In what seemed like only a minute later, I was awoken by drips of water plopping on my cheeks and nose. Tequila was fidgeting and growling. I opened the netting so he could get out, thinking he must need to relieve himself, but he nuzzled further into my belly.

The storm was raging and I couldn't get back to sleep, so I lay listening to the crack of boulders tumbling along the river, and the distant crack, crash and thud of falling trees. The tarp and flysheet were so soaked through they'd become sieves. Rain sopped through, splashing down and soaking my blanket and clothes. Tequila, clearly unhappy about the soggy conditions, whined throughout the rest of the night.

At five in the morning, I couldn't stand it anymore. I got up to stamp and pace, to kick-start the circulation back into my cold, damp limbs. Tequila followed suit, stopping every now and then to shiver and shake out his mop-like hair. By first light I'd decided to break camp and head back to Tena. The river was too high to work and I was wet and miserable. Besides, I'd made more than enough gold to pay for a few days of food, drink and lodgings and to buy the supplies for my next trip.

Unfortunately, the beach, my way in and out of camp, had disappeared beneath the rapids, meaning I'd have to cut a new path up to the main trail if I wanted to get out. I hid what I was able to and loaded everything else into my backpack. Bent under the sixty-pound weight and soaked to the skin, I began hacking my way up the bank behind my camp.

My feet slipped on the slick mud every time I swung the machete, making the steep climb out even harder. When I reached the trail, I was already exhausted. I didn't want to stop, though. I had to keep moving to stay warm.

The trail had turned into a dirty-orange stream, gurgling and bubbling between the dense foliage. I tramped on, carrying Tequila in my arms. Every so often, I'd sink down to my knees in a muddy puddle. It was like quicksand, sucking and gripping my legs so hard I'd have to use overhanging branches or stout bushes to pull myself out. During these tense moments, Tequila would jump from my arms and yelp encouragement from beneath the nearest vegetation while I wriggled and thrashed my way out.

On the higher ground where the main-trail ran, the storm must have beat through with the force of a hurricane; everything was whipped up and reeling. Leaves and fallen branches littered the ground. Entire trees had been uprooted. Some lay across the trail, damming the flow of water and making wide muddy pools, and high, slippery, wet leaf and moss-covered obstacles for me to negotiate. I had to push Tequila up onto the top of the gargantuan toppled trunks before struggling over them myself.

We reached the creek where Tequila had learned to cross on stepping-stones. The rocks we'd used were no longer there and a wide gush of water blocked our path. I picked Tequila up and tossed him into a stand of grass on the other side before wading through the thigh-high cascade, fighting the dragging current with every step.

The second creek we came to was even worse. This one was double the width of the first and tumbled through a steep-sided ravine. Grabbing at rocks and roots, I scrabbled down as close to the water as I dared. Tequila followed, skidding and slipping behind me. I took off my backpack, trimmed a fallen branch, and used it to test the water's depth. Tequila looked on, perplexed. The stick didn't reach the bottom. It caught in the current and pulled away to join the flow to the wide river below. I let it go and watched it race away.

Turning, I grabbed Tequila and chucked him across for the second time before he could realize what I was doing. He landed on the mud bank with a splat and a yelp, and tumbled back toward the rushing water. I held my breath, thinking of Chico plummeting from the rock into the icy Alaskan river all those years ago. Tequila was much smaller and less river-savvy than Chico. I was frightened he'd be swept away and never seen again.

Somehow, though, he managed to halt his momentum. Legs splayed, he barked at me from across the creek.

"I know, boy. I'm sorry. If it makes you feel any better it's my turn now."

Tequila wagged his tail and made a start toward the water's edge.

"No!" I yelled. "Stay there!"

He looked unsure, but did as he was told. I lowered myself into the foaming current and by the time I found a footing, the water was chest height and pushing hard. I reached for my backpack and hoisted it onto my shoulder. Leaning into the flow, I inched forward, feet grappling with shifting stones. I kept focus by keeping my eyes on Tequila, who was willing me on, his own dark eyes pleading with me to make it across.

As I got closer, his tail flicked side to side. Slow at first, the momentum built into frenzy by the time I reached the bank and dropped the backpack beside him. I climbed out of the water, and he did his funny little circle prance around my legs. The two of us then scrambled up the bank and trudged onward.

By now, we were getting closer to the park entrance, and all I could think of was getting a warm meal and a cold beer from the women who ran the shack-cafés next to the pools. It was Saturday and they'd be there early, cooking breakfasts and prepping lunch. With any luck the sun would come out as well and I'd be able to dry off a bit before hitching a ride back to Tena.

The pools were a series of swimming holes fed by mini waterfalls, cascading all the way down to the main river. They were popular with locals and tourists and were full of people on the weekends. With all the comings and goings, it was usually easy to find someone willing to take me back to town.

I could hear the thunder of the waterfalls a full fifteen minutes before I reached the pools. By the time I arrived, the sound was deafening. The pools were a roiling mass, violent and alive, splashing and soaking the boards of the flimsy rope bridge I had to cross. Tequila cowered back, and I felt like doing the same.

"Come on, boy," I said, crouching and reaching out for him.

Tequila looked left and right, searching for an escape route should I make another grab at him. I took a piece of three-day-old dried fish from my pocket and gnawed on the edge before offering it to him. I was so hungry, the moldy fish-leather tasted good to me. Tequila craned his neck to sniff. He took the fish, his eyes on me all the time.

When he started to chew, I seized the moment and scooped him up, one handed. Holding him to my chest, I felt him chomp down his mouthful as I started for the bridge. I smiled. If this was to be his last meal, there was no way he wasn't going to finish it. I gripped the rope handrail with my free hand and moved as fast as I dared over the slippery boards.

My heart dropped as soon as we reached the other side and I put Tequila down. The shacks were all locked up. The place deserted. I called out but no one answered. Nobody was there. A wave of fatigue washed over me, and I felt like crying. I'd left what snacks I had behind - stowed in a Tupperware container back at the camp - because I was sure I'd have a hot breakfast waiting for me once I reached the pools.

I took a sip of water and pulled myself together. Of course there was nobody around; last night's storm had been brutal and it was still overcast and drizzling rain. With leaden legs, I trekked on out of the park and onto the road.

I knew there was a village a couple miles away. I'd already walked three miles through the wild assault course of the jungle, two more on a wide muddy road should be easy by comparison. Besides, I figured I'd probably get a phone signal further up and be able to call a taxi.

As soon as I exited the park, I saw a sign I'd never noticed before jutting out at the side of the road. It read, 'Peligro!', *Danger!*. Just beyond, the cliff bit into the road in a ragged semi-circle of erosion, reducing the width there to one lane.

I peered over the edge and was hit by a sudden attack of vertigo. I was so lightheaded I was afraid I might topple over and plummet to my death in the rushing river far below. I staggered away from the edge and took a few deep breaths, cursing the storm and the jungle and my empty stomach.

Before setting off again, I rummaged in my pocket for my cellphone. I found it and hit the power button. The battery was dead. I must have left it on and let it drain out.

"Damn it!" I shouted.

Tequila cocked his head at me. When I said nothing more, he trotted on ahead. Thoroughly miserable, I followed him up the hill. There were some cabin houses half way up. I figured maybe I'd find some sign of life there. Storm or not, it *was* the weekend after all.

When we reached the houses, I leaned against the gate and hollered though the bars. While awaiting a response, I studied the four cabins. They were raised up on stilts and made of solid hardwood. I imagined the views of the jungle from inside must be stunning.

Beside me, Tequila gave a yip. He was looking up the hill at two people rolling down toward us on a bicycle. One of them was sat on the seat, the other was standing on pegs either side of the back wheel. I thought they were kids at first, but as they drew closer I could see they were grown men in their early twenties.

They were indigenous and could have passed for Moipa's and Kimo's elder brothers, but for the fact they wore their hair a lot shorter than the Huaorani. Both were broad shouldered and well-muscled. One was slightly slimmer, the other was a full-on titan. Neither was wearing a shirt.

I waved at them. They waved back and continued freewheeling through the spraying mud until they came to a stop a few yards away.

"Buenos Días," said Slim, from the back pegs.

The Titan hunched forward over the handlebars. He nodded and smiled a greeting to me.

"Hola," I said, smiling back. I held up my cellphone. "No teléfono. Necesito taxi a Tena."

They had a quick conference in a native language. It could have been Huao or Quichua, I couldn't tell which. The Titan took a cellphone from his pocket and passed it to Slim who held it up high, apparently trying for a signal.

"No," he said, shaking his head and passing the cellphone back. He pointed up the hill. "Hay arriba." He was telling me they could get signal further up the road.

"Si?" I asked, miming talking into a phone, and gesturing ahead.

"Si," they chorused, nodding and grinning like loons.

Slim got off while The Titan turned the bike around. Then he got back on and the two of them set off up the hill. Slim looked over his shoulder and waved for me to follow.

The two men stopped every now and then to check for a signal. Each time they did, Slim wandered over to the cliff edge and looked down. On the bike, they were a lot faster going uphill than me. They drew further and further ahead until I lost sight of them around a bend.

I kept a decent pace but didn't hurry. I was too weary, and I figured they would ride back and let me know if they got hold of a taxi. When I reached the bend at the crest of the hill, I saw the guys standing together waiting for me. The bike was lying on the ground at the side of the road. They were both concentrating on the cellphone in the bigger guy's hand.

As I approached, he raised it to his ear. "Si," he said, ushering me onward. He walked forward to meet me. "Hola? Taxi? Taxi?"

Tequila growled and I stopped in my tracks. There was a subtle shift in the atmosphere. Tequila sensed the growing intensity, the menace, and so did I, but far too late to do anything about it.

The two men were now five feet away. The phone had disappeared and I had no doubt I was in danger. Fear made my empty stomach squirm. Tequila barked a final warning to me, but there was nothing I could do. I wouldn't stand a chance against these guys in the best of circumstances, let alone while exhausted, hungry, and carrying a sixty-pound backpack.

They rushed me. Slim threw a punch at the last moment, and I instinctively raised my arm to block it. As I did, The Titan barreled into my exposed ribs. I heard a crack and gasped in pain. The force of the blow knocked me off balance, and the weight of my pack pulled me over. Slim's fist arced through the air, missing me by a mile, as I toppled backward landing with a splat, belly up in the mud.

Pinned like a turtle, arms and legs waving in the air, I must have looked ridiculous. The men jeered and laughed at my struggles to get upright. Then Slim jumped on my chest. His weight pressed down on my broken rib, making me whimper. I couldn't understand why they were attacking me. If they'd wanted money, why hadn't they asked for it?

Slim clamped his hands around my neck and began to squeeze. I tried to throw punches at his face, but the backpack straps restricted my movement and the blows lost all impetus. I tugged and pounded at his arms, but it was no use. They were set like stone.

"Don't kill me!" I begged, my voice a rasping whisper. "Please! Take it! Take my stuff. Just don't kill me!" I was so frightened now, I was sobbing. I didn't care though. All I cared about was not dying, and I had to let them know it.

The big guy loomed into sight and grabbed my shoulder. He hammered his fist into my right eye, over and over. Silenced now by the hands around my neck, I pleaded with my left eye, first to one then the other. But they didn't see me. They had no compassion. No remorse. No pity. Even worse, there was no anger or enjoyment either. It was all mechanical. I was nothing to them.

My vision blurred to soft crimson dots. I screamed, the sound coming out a whiny croak. I didn't stop. I couldn't stop. It was a primal cry of agony and indignant terror. More than anything, it was a plea for them to see I was still a living human being before it was too late.

The hands at my throat closed even harder, choking away all noise aside for the meaty smack of The Titan's fist pummeling my face and Tequila's incessant barking and growling. At least I thought it must be Tequila. He sounded like a different dog.

I gulped a tiny breath and bit my tongue. Hot rusty liquid filled my crushed windpipe. Lights danced in my eyes. All around was blackness, so deep and dark that I knew if I let myself fall into it, I'd never leave. This was it. They were not going to stop. They wanted to kill me and I was allowing them do it.

The shame of that admission boiled into a rage. Every disappointment, every miserable defeat I'd ever experienced welled up inside me. Like molten lava pushing through the thick crust of a volcano, an explosion of indignant fury rushed through my body, fizzing in my limbs.

I shoved The Titan away and heaved Slim off me. Sucking in air that seared my throat and stabbed my ribs, I renewed my struggle to get to my feet. The Titan hadn't been pushed far and was already coming back toward me again. I raised my legs to keep him at bay, and he lashed out a vicious kick that connected with my right kneecap and made me howl.

While his foot was still in the air, Tequila dashed at his standing leg and bit down hard on his ankle. The Titan grimaced and stamped down on the tiny dog. Tequila yipped and whined and rolled away.

"You bastard," I hissed. "He's a dog for fuck's sake."

The Titan's hands shot out and grabbed my calves. His grip was so strong that I couldn't move. He yanked and the pain in my knee lit up like wildfire, making me sweat and pant and weep all at once. He was dragging me now, across the muddy road toward the cliff edge. Slim was there at my chest, fumbling with my backpack straps, trying to get it off.

I helped, cursing myself for not thinking to unclip them earlier. Once I was free of the backpack I fought back again, kicking, and lashing out with my scrawny arms, my pinprick fists. The big guy dropped my legs and grabbed my shirt, pulling me up and groping at my side.

I didn't know what he was doing until I remembered what was there: my machete. It had been hidden under the backpack until now. He pulled it from the sheath and shoved me back into the mud.

Holding me down with one hand, he raised it over his head, ready to chop down. I crossed my arms in front of my face. This bastard was going to kill me at the side of the road with my own machete. I couldn't understand why nobody had come down the road yet. It was Saturday. Somebody had to come. I closed my eyes and prayed for a car to round the bend and see us.

No one came, but Slim shouted something and The Titan held off. The machete hung in the air, sharp and ready, while Slim pointed at the mud all around me. He then jabbed his finger toward the cliff edge. He repeated something, a phrase I didn't understand, over and over. The Titan nodded and threw the machete across the road.

I turned my head to follow its arc. Fresh globs of blood from my mashed-up face formed a dotted trail along the ground. I realized the reason they hadn't used the machete was because they didn't want an obvious pool of blood at the scene of the crime. I felt sick, knowing they meant to murder me and hide my body.

A part of me wanted them to get it over with, to kill me already. I didn't want to give up, but I felt so hopeless, so drowsy and in pain. I

Falling to Fly

wanted so much for it all to stop. When they went to grab me again, though, survival instinct kicked in. I twisted and squirmed and managed to scramble to my feet at last.

I felt light as air without the backpack and took off running. After a couple of steps my stiff right leg buckled. My knee was bad. My right eye was closing up and I could hardly breathe, but I hobbled on as fast as I could. Tequila was beside me now, barking and wagging his tail, urging me on. I noticed he was limping too.

Something bashed into the blind side of my head and I fell. My mouth filled with blood and mud. My temple throbbed and I retched. Bile burned in my ruined throat.

The men seized me again. The big guy with his anaconda grip had my legs. Slim held my arms. They ran together to the cliff edge and hurled me up into the air.

I pumped my arms and legs, vainly trying to gain purchase on nothingness, thinking I could stay aloft; such was the depth of my shock. Everything was beautiful and suspended in time - the never-ending lushness of the jungle, the wet umber road, the rolling cappuccino river, even Slim and The Titan, standing still as statues, breathing hard. The rise and fall of their chests brought breath and motion back to the world and I fell fast. Crashing, rolling, and dropping, again and again, I bounced and smashed into rocks and bushes and earth.

Adrenaline stretched time and between each shattering impact I was somehow able to gather my thoughts and make a plan. I remembered the fall from my childhood through the massive pecan tree and knew what I had to do. I kept my head safe and grabbed at everything and anything I could. The problem was nothing held. My momentum was too strong, the cliff too steep. Blood and dirt and stone kicked up all around. The river was getting closer and I couldn't help thinking how the Jatanyacu would get to devour me after all.

Then it happened: a clump of vines rose up and knocked me sideways into another cluster that sent me tumbling away from my initial trajectory. The vines appeared suddenly and from nowhere, catching and throwing my beat-up body ever leftward. They were slowing me down too, but not nearly enough.

I skipped off the edge of a sixty-foot drop-off head first toward a tumble of rocks. Then I heard a crack, like a tree breaking in a storm, and a vine - thick and long and snapping like a bullwhip - unfurled from above and wrapped itself around my calf. It jerked me up so quickly, my leg felt as if it were being yanked from its socket. I swung across and down. Away from the piled-up rocks. The vine released me and I dropped onto a slip of mud.

Winded, but still living, I continued to hurtle toward the river. A dark, bushy guayusa tree was in front of me now. I lunged for its fat trunk and caught hold. Slipping and fighting gravity and motion, I clung on with all my might.

Then it was over. I was six feet above the river and alive. I'd survived the fall. I clambered onto the near-horizontal trunk and kissed it, then reached for a vine. A thunderbolt spasm of pain pulsed up and down my leg, making me sweat and wince and pant.

My thoughts and emotions rushed. Details of the attack and fall raced through my mind, every moment as sharp and immediate as the pain racking my body. The moving vines. The vine that reached out and caught me. They were real, yet they made no logical sense. I didn't know what to make of it all, but I was so grateful.

A boulder crashed through the leaves of the guayusa tree, spraying muddy water all over me and returning my attention to my immediate predicament. My bewildered speculation about the vines could wait. I was still far from safe. The river was rising and tearing along, tossing huge boulders that spun and smashed and splashed ever onward. I was perched on a tree whose roots were exposed and laddered, splaying and thinning the closer they got to the earth that held them. These twigs were all that kept the tree, and me, from being swept away into the hungry waters below.

Above, was the high, crumbling cliff-face and my would-be-murderers. At the thought of them, my heart hammered and I pressed my body against the vertical mud bank. I couldn't see them, but I was paranoid they might see me. What would they do if they knew I was still alive?

All around I saw debris; water bottles and hiking kit, a walking pole, a broken machete, a map, torn, wet, flapping. There was clothing too, and the sickly-sweet stench of rotting flesh. I gagged. A flash of black above caught my eye, and I turned to see my backpack arc down into the river with a splash. My machete and jacket followed. The latter fluttering like a great wounded crow, before churning away downstream.

I tried not to think about Tequila up there alone with those monsters. I hoped he'd seen what had happened to me and ran away as fast as he could.

He hadn't.

Somehow, over the thunder of the river, I heard his shrill death cry. The high, strangled whine of pure terror cut through me and broke my heart. It cut off abruptly. My friend was gone. I bowed my head and wept.

Jason Pednault   Matheu DeSilva

Sisu

It started to rain again. Not heavily, but enough to bring me back to the moment. I wiped my left eye. The right was swollen and closed over, too tender to touch. My vision was more blurred than usual, even in the one eye that worked. Every part of my body hurt, from my broken ribs to my battered, throbbing head. My right knee wouldn't bend, and any movement in my right hip set off a wave of pain so severe it made me nauseous. I could barely breathe as well. My windpipe was crushed, scratchy and parched.

I considered relieving my thirst by drinking from one of the water bottles strewn around, but decided against it, imagining my attackers might have filled them with poisoned water to catch me out. I knew I was being paranoid, but figured after the morning I'd had it was better to believe anything was possible than not.

I had an idea to gather the water bottles and old rags and tie them together into a distinctive marker. I could climb along the tree and hang it over the water, then wait for someone to see it and come investigate. Another ridiculous thought. Nobody ever came down this stretch of the Jatanyacu. The rafters put in downriver, and there was nothing upstream but the wild, uninhabited Llanganates Mountains.

Besides, the river was running faster than I'd ever seen, with the possible exception of the night my Vilcabamba crew got flooded out. And I hadn't so much seen it as felt it then. Rocks clattered and whole trees pitched like sticks in the ripping, rising, chaos. Setting a marker would be pointless. The only way someone would see it was if the river swelled high enough to carry it and the tree I was sitting on all the way down to Misahualli. With the rain growing heavier now, that scenario seemed more and more likely and not the least bit comforting.

The only way out was to climb the hundred-and-fifty-foot vertical wall of slick mud, slippery stones, scraggly bushes, and reedy vines in the rain with no ropes, in an area notorious for sudden and unpredictable mudslides. Not something anyone would choose to do in the best of situations. Half blind, with busted bones and a dislocated knee it would be impossible.

Surviving the attack and the fall, only to die of exposure or drowning at the bottom of the cliff, was an irony so perverse I felt like screaming and shaking my fists at the gods. I kept quiet though. There was no way I wanted my attackers to realize I was still alive. Being crushed under a boulder dropped from above might be a quicker way to go, but if I was going to die, I'd rather it be on my own terms and in my own time.

Filled with despair, I began thinking of my family and friends. They'd never know what had happened to me. The rest of their lives would be spent wondering. I pictured my mother weeping at home, while my dad dragged Mary to Ecuador to search for me. I knew he'd be unwilling to let me go until he knew for sure I wasn't coming back. Tears filled my eyes again. I was going to put them through years of hurt, uncertainty and grief.

Flashes of memory played through my mind. Ordinary moments rendered poignant and heart-wrenching now I would never see any of them again. I smiled at my childhood self, forever climbing trees while my sister called through the woods, trying to find me and bring me home for supper.

Monkey Boy. I was always the climber. In some ways my whole life had been about climbing. There's nothing I'd loved more. Not gold, not snakes, not a woman. Not yet, anyway. Climbing had always been my thing. I gazed, one-eyed, up at the cliff-face. I couldn't even see the top. But still I only had one thought in mind:

Let it be on my own terms.

*

I still had some residue of adrenaline left in my bloodstream, and now I'd decided to climb, I wanted to make the most of it. If I didn't, I wouldn't make it; with each passing moment my beat-up body was getting stiffer, more seized-up.

When I stood, I felt an uncomfortable lump in my front pocket. It was the old camera film can where I kept my gold. I took it out and wedged it into a fork in the tree. As a kind of offering. I made a secret promise, with the tree and the roaring river my only witnesses, and then turned to the cliff-face.

I wanted to climb to the right. The bushes were denser in that direction so I figured I'd have more to hold on to. It would also take me further from where my attackers had thrown me over. The problem was, I couldn't see out my right eye and I didn't want to have to keep turning my head around on my damaged neck.

Left it was then.

I pulled at the base of a stumpy shrub. It was strong enough to support my weight. I went for it and pulled myself up and away from the relative safety of the tree. The bush held, and I kicked my foot into the mud, searching for something solid to use as a foothold. I found a rock. So far, so good. I pushed myself up and reached over the bush and felt around for a new handhold. Another rock. This one was jagged and flat on top. I tugged and it stayed in place. I shifted my weight and then dragged my body over the scratchy, snapping twigs of the bush.

It was slow, painful progress, but I was making it. Always moving to the left and climbing higher all the time. Sometimes I was able to test a handhold or foothold in advance, but other times I had to leap for one and trust that it would hold. I kept going like this for about forty-five minutes. Up and left. Left and up.

I arrived at a large scraped out area of erosion; the site of a fresh landslide. There were no plants, no rocks. Nothing but glossy mud in a slick patch, twenty feet across. Forty feet down, there was a cluster of boulders and sharp snapped tree trunks and mud. I cocked my head to see above. Nothing there. There was nowhere to go. My path had come to a dead-end.

I inched back the way I'd come. It was more difficult moving to the right, just as I'd known it would be. I had to turn my head all the way around so my left eye could see where to place my hands and feet. The movement made my crushed neck muscles tighten, and my sand paper throat pop and spasm. The route I'd taken was still fresh in my mind, and after a long hour's climb I was able to make it back to the tree.

The river was higher now. The white tips of the rapids battering and splashing the streaming foliage of the tree's lower-hanging branches. I had no time to rest and no choice but to try climbing to the right. Ignoring the cramping in my neck, and everywhere else for that matter, I started again.

There were definitely more shrubs this side, and they were a lot closer together. There were also scrawny trees, their tenacious roots digging into the cliff-face. I wasn't a rock climber, but all my years of bouldering and climbing trees had honed my technique to the point where I found I could compensate for my lack of mobility - particularly in my right leg - by use of arm strength, footwork and weight distribution. I leaned, using my left leg muscles to keep my body suspended in place while grasping for handholds. I switched feet and leapt, twisting, grabbing, pulling until my left foot found purchase again. The whole of this part of the ascent was like some exhilarating dance.

I reached an enormous, spiky bush and had to switch back to make my way up and around it. Six feet above the bush, a narrow mudslide stopped me in my tracks once again. There were wide rocks jutting out and away to my left, but there was no more foliage for at least twenty feet in any direction.

I had no idea how deeply the rocks were embedded into the cliff-face and there was no way of knowing if they would support my weight until I'd reached them. Once I'd committed, I'd have to hope they held. At least it had stopped raining now and I'd be traveling leftward and be able to see where I was going.

The first rock was strong and broad, but it was also rounded and wet. I fumbled along, my hands slipping and slapping as my legs dangled and jarred and throbbed. My left hand lunged for the next rock, a square block that looked as if it had been quarried and cut for a step. It was solid and I was able to take a breath before moving on. I swung and grabbed and scurried, hands skipping onward and upward, rock to rock, in quick succession, eighty feet up in the air. All the way along these bulbous, slippery, tricky monkey bars, I kept my eye fixed on the solid-looking tree just beyond the last stone; a point where I could rest and gather myself again.

Whenever I grabbed a tree or bush, I always did so at the base, where it was strongest and I was less likely to put too much strain on the root system. This tree was thick through the trunk, though, and at a reassuringly upright angle, so I lunged for it, not caring where I caught hold, just glad to be no longer hanging from the precarious rocks.

The tree was half rotted. It cracked and snapped off in my hands, bashing into my shoulder as its weight pulled it downward, plummeting all the way to the river. I fell with it, and a rasping scream hissed in my throat. Ten feet down, my back struck a supple sapling and my descent was halted. Instead of falling, I bobbed in midair, unable and unwilling to move.

Splayed horizontal, I was balanced on what felt like an impossibly thin tree. It ran beneath my seized-up right leg and under my back to the inside of my left shoulder blade, with the top poking out alongside my head. Leaves, frail and fluttering, tickled my left ear. I dared not move. My ruined neck muscles strained and constricted with the effort of keeping my head up.

The sheared off stump of the tree I'd broken seemed a million miles above me now. There were a few tufted scrubs between it and me. Not many, but maybe enough to get me back up there if I could maneuver off the sapling without falling. I twisted my head to stretch my neck and caught a glimpse of what was behind me: eighty feet of nothing, then the river.

The big tree I'd fallen from was already long gone. Dashed to pieces and whisked away. The clamor and constant motion of the river made me lightheaded. The clacking of boulders jarred in my skull.

"Holy shit," I whispered, staring up now, into the gunmetal sky.

If I'd fallen an inch further left or right I'd be dead already. Instead of feeling fortunate, though, waves of tiredness and hopelessness washed over me. My muscles were shutting down and stiffening and my blood congealing in my veins like some kind of premature rigor mortis.

While moving, I'd been able to ignore my injuries to an extent. Now I was forced to be still, every sinew and joint in my body sizzled in agony. I didn't want to move and risk dislodging the small tree, but the longer I stayed where I was the more likely it was I'd lose balance or consciousness and fall anyway.

Slow and steady, my left hand reached across my thigh and down between my trembling legs to grip the tree. Holding tight, I brought my right hand over and placed it further down the trunk. Hand over hand, I pulled myself upright until I was straddling the solid little tree. At last I could see what had, and was continuing to, save my life. The sapling was three inches wide where I sat, and flared to no more than five at the base. It defied the laws of physics.

The river continued to roil and rumble far below, but now I was moving again it no longer had the same disorienting effect. I was back in the survivor's mindset - all my senses enhanced and fully focused on getting to the scrubby bushes overhead. The pain still racked my body, but I was able to put it out of my mind. Adrenaline really is amazing stuff.

I shimmied my way down the sapling until my face was up against the damp mud of the cliff. My movement had shaken and displaced some of the earth around the sapling's roots, but still it held firm. I took a breath and watched the dirt shift and seethe and come alive, revealing a swarm of black ants.

They stormed onto my hands and up my arms. I flinched away and swiped at them, my frantic flapping almost making me topple from my perch. The ants were relentless. Rushing up the bark, they attacked me in earnest, teeming over me and burrowing into my clothes. They nipped and gnawed my bruised skin until I was covered in a rash of itchy bites.

There was nothing to do but make a run for it. In one smooth motion, I brought my left knee up to my chest, hooked my foot over the sapling and pushed up. I felt it give under my weight and motion, but by now it didn't matter. I'd already grabbed the nearest tuft of scrub-grass and was hoisting myself upward, away from my second arboreal savior of the day, and, more importantly, from the ants.

The tufts were set to perfection. Any further away from each other and I wouldn't have been able to reach them. I hurried upward, until I reached the torn off stump. There was no other way past. I had to climb over it.

Creaking and making sharp popping sounds, the stump held just long enough for me to scramble onto a narrow shelf of rock above it before upending in a spray of dirt and a hiss of scattering stones. It dropped a few feet and then hung, suspended and swinging by one thick root like a grotesque mutant spider. With a twang, it plunged further still until it came to rest against the resolute sapling. The small tree was having a busy day.

Before moving on, I began the agonizing process of beating and rubbing as many of the ants as I could out of my clothing. Once I finished, I took a moment to look out across the jungle. Everything glistened and steamed. The sun was pushing through, creating an alchemical transformation in the clouds - base lead into gold.

The persistent sun lit a spark of hope in me, too. I'd made it over halfway up and was still alive. If the cliff dried out, the way up would surely get easier. I started to think maybe I could make it. This sudden optimism was short-lived, however. The climbing got harder and scrappier the higher I went, and the hazy sun was now baking my already dehydrated brain.

I struggled and sweated. Moving in slow motion from rock to bush to vine to tree. Exhaustion was setting in, and my throat was so dry. Every breath I took was parched and scratchy and sounded like a wheezing death-rattle. The cliff appeared to be stretching out, mocking my efforts to conquer it. Ants and a myriad of other biting things were hidden under every handhold. My hands tingled with stings and thorn scratches. Things were getting worse, not better.

I came to another gouged out site of erosion. There was no way up or across in any direction. I was stuck again. What made things worse, ten feet away, across the sheer muddy scoop of the landslide, there was a gently sloping, grass-covered shelf. As close as it was, this resting place may as well have been on the moon. Another ten feet away, at the top of the landslide was a vast and sinister-looking tree. It was leaning forward and had a maze of freshly exposed roots stretching as wide as its bare branches.

I clung to the base of a brittle bush and tried to figure out what to do next. As I hung there pondering, my eyes were constantly drawn back to the big tree. As if it might offer some answer. I couldn't think what it might be, as it was too far away to be of any real use.

Or was it?

The tree was wide and tall, and the roots I could see mirrored the branches in size and span before they disappeared away into the ground. As if divinely guided, I dug my fingers into the damp mud beside me where the earth had so recently been sloughed away. It didn't take long for me to find what I was looking for: the furthest extent of the mirror image. I kept scratching away until I'd exposed a dense curl of root, hard as stone and thick as my thigh.

I snapped a twig off the bush I was holding and used it to mark the location of the root. I could see it well enough now, but once I started climbing again it would be rendered invisible, set back into the cliff-face as it was. Once I collected a decent sized bundle of twigs and wedged them into my belt, I began excavating again, this time as high up as I could reach. With my initial handholds prepared, I left the bush and began the slow process of crossing the slimy mud wall.

It was arduous work. Clinging, one-handed, while scraping at the mud - searching for the next root - with the other. Sometimes finding one and sometimes not. The twigs helped enormously, providing me with a way to locate much-needed handholds and footholds at a glance.

As if the painstaking progression wasn't hard enough, half way across I had to deal with a parade of inch-long orange fire ants. Their sharp stinging nips, pinched and burned my fingers and wrists, making it even harder to hang on. In the end it took me close to forty minutes to cross an area ten feet wide. When I reached the grassy incline on the far side, I collapsed and tried to catch my breath. Once I laid down, I wasn't sure if I'd ever have the strength to move again.

As I lay there, I noticed some odd-looking trees above and across from where I rested. They were an explosion of branches and leaves, like the crown of a tree, but with no discernable trunks beneath. Intrigued, I scrabbled up toward them. The closer I got, the bigger I realized these strange trees were. They were easily as big as the colossus whose roots I'd used to traverse the landslide.

I came up around and discovered they were, in fact, quite ordinary trees. They'd just been uprooted and were now hanging upside down from their roots. There were four of them, staggered one atop the other, at an angle just oblique enough to have arrested their slide down the cliff and into the river.

I clambered through the branches of the lowest tree and scaled the fifteen feet to its root ball. Here, I skipped to the left into the next tree to repeat the process again. The four trees made an easy ladder for me and I was able to ascend nearly fifty feet in under ten minutes. While climbing, I heard something that gave me a burst of energy and hope: a motorcycle passing by on the road above, heading toward the pools. The warm sun was bringing people out at last.

The last cluster of roots was a challenge. They were sharp and tangled, and took another quarter-hour to negotiate. When I got past them, I felt disoriented. Something wasn't quite right, but for a few long seconds I couldn't figure out what it was. Then I realized: the cliff was gone. After climbing so long, it didn't make sense that it could all come to an end.

I turned and stared down at the river, so small and distant now. I was alive and at the top. By some miracle, or rather by a series of miracles and no small amount of bloody-minded perseverance, I'd made it. I grinned and would have danced for joy if I didn't think I might topple back down the cliff-face if I did.

There was a thicket of dense bushes in front of me. I crawled up into them and pushed through until I got to the edge of the road. I was too afraid to step out into the open in case my attackers were still there, so I lay hidden in the undergrowth, wondering what to do next.

I had three options: go to the village, go back to the pools, or stay where I was. The village or the pools both posed the risk of meeting my attackers again, but I'd come too far and was too hungry and thirsty and beat-up and exhausted to stay hidden under the bush forever. Besides, the murderous bastards had probably run for it once they believed me to be dead and gone.

I decided on the pools. They were closer than the village, and I knew somebody was there now. I'd heard their motorcycle. If whoever was there didn't look friendly, I could always sneak in and hide in one of the kitchen shacks and wait for the cooks to arrive. They knew me and would be able to help.

I scrabbled out, climbed to my feet and limped off down the hill, staying close to the foliage at the right side of the road in case I had to dive for cover. My right leg was fused straight. Pain jolted with every hobbling step. The combination of the heat and humidity, coupled with chronic pain and feeling exposed was making me drip with sweat. I was amazed I still had any liquid left in my body.

When I reached the site of my attack, I froze, unable to walk on. As irrational as it was, I could still feel the presence of my attackers, and of poor, brave Tequila and his final struggle. There was no sign of what had happened to us beyond a few scuffmarks in the sand. The bastards had covered their tracks well. My father would never have found me.

I picked up two fist-sized rocks and hurried on. Although I hoped with all my heart the guys were long gone, I couldn't help fantasizing about what I'd do to them with the rocks if they weren't. I oscillated between smashing their skulls and some long, drawn-out torture. I'd never hated anyone as much as I hated them. As far as I was concerned, they deserved to suffer and die, preferably at my hand. I realized I was shaking. From anger or fear, I couldn't tell.

The motorcycle I'd heard was parked at the entrance of the park by the little wooden gate. I didn't see a bicycle, so was pretty sure my attackers weren't there. Still, I opened the gate as quietly as I could. The motorcycle might mean more guys. Maybe friends of the two who tried to kill me.

I crept along in my gimpy shuffle and rounded a corner. On a picnic table, twenty feet away, eating lunch, was a family; a woman, a young boy and a toddler. I was so relieved, I fell down on the ground and wept. Through my tears, I saw the family rise and come over. The boy, who couldn't have been more than fourteen years old, put his arms around my trembling shoulders and held me while I sobbed.

"Dos hombres," I croaked, *two men*, and gestured toward the cliff.

The mother nodded and the boy patted my back. The toddler hid behind his mother's leg. I didn't blame him - I must have looked like a monster. My face was smashed and swollen, and I was covered in itchy red ant bites.

They invited me to their table and gave me food and soda. As hungry and thirsty as I was, I could only nibble at a small piece of chicken and take a few sips of the fizzy, rose-colored drink before I felt too nauseous to go on.

After a brief discussion between mother and son where I heard the word police mentioned several times, the boy led me back to the motorcycle while his mother stayed behind with the toddler. The boy started the machine up and gestured for me to climb on the back. It was difficult, but after a few minutes of stiff-legged agony, I managed it.

As we bumped up the road, it occurred to me the boy and his family were indigenous, and possibly from the same tribe, or village, as my attackers. My heart raced and I felt sick, and I had to fight the irrational urge to fling myself off the bike. Forcing myself to calm down, and gritting my teeth against the painful jarring, I tried to think straight. These people had shown me nothing but kindness. The boy was taking me to the police. I had no reason to be afraid anymore.

The village was further than I'd thought. More like four miles than two, and the ride there was a bone-juddering ordeal. When we arrived, we found a lone policeman working on his motorcycle outside the small concrete building that served as the local police station. He was wearing a filthy soccer shirt over uniform pants, and a utility belt that held a radio, three small cans of pepper spray and a scratched-up pistol.

At the sight of us, the cop stood and wiped his hands with a rag. He nodded and glanced occasionally in my direction while the boy explained what he understood of what had happened to me. When the boy had finished, the cop turned to me.

"Dos hombres?" he asked.

"Si," I said, grimacing as I slid of the back of the boy's bike.

A crowd of grim-faced indigenous had gathered. Most remained silent, but some jeered at me. I scanned their faces to see if I could see either of my attackers. They weren't there. I still held a rock in my hand, and felt braver with the policeman standing behind me. I wished they were there. I wanted them to show their faces.

"Ven," said the cop, waving me over. "Vamos al hospital."

I said thank you and goodbye to the boy then walked over to the policeman. He'd taken off his dirty t-shirt and put the other half of his uniform on already and was in the process of securing the strap of his motorcycle helmet. Leaving the wrenches strewn on the ground, he kicked started the engine and pulled forward so I could get on the bike.

As I struggled to climb on with my busted leg, I could hear chuckling break out among the crowd. The cop laughed, too. I guessed he had to. He lived with these people. The crowd parted so we could pass, and once I was settled, we took off.

As soon as we were out of the village, the cop accelerated up the muddy lane. The back wheel of the motorcycle slid and spun as we ploughed through puddles. I hadn't been given a helmet and was terrified we were going to crash. I hoped he'd actually finished his tinkering when the boy and I had arrived, and that all the nuts and bolts were back where they were supposed to be.

His riding got worse when we reached the paved road to Tena and he really opened the machine up. I squinted over his shoulder with my good eye. The speedometer read over a hundred kilometers per hour. All I could think about was dying on the road. After everything I'd survived, this idiot was going to kill me on his motorcycle.

When we arrived at the hospital, I was so eager to get off the bike that I fell onto the sidewalk. The cop helped me up and half carried me into the emergency room. Fortunately, one of the doctors there spoke some English. He gave me painkillers and explained there wasn't room to admit me, and that the x-ray machine was broken.

"Come back tomorrow for x-ray," he said, frowning at my knee. "Tomorrow is good. You want I translate for police what happen now?"

"Maybe tomorrow," I said. I didn't think either his English or my energy was up to recounting everything I'd been through right then.

It was now early evening and the cop took me to back to my hostel. Pablo and Carmen weren't there, but their daughter Belén was. Her eyes grew wide at the sight of me, and she shot a bunch of questions at the policeman. He shrugged and muttered a few words in reply. The only one I understood was 'mañana', *tomorrow*.

Belén gave me the key to a ground floor room, and the cop helped me get settled in before he left. I drank some water and took a couple more painkillers then lay down on the bed, my head spinning. I was shaking and every inch of my body hurt. Thankfully, the pills were strong and I was soon feeling numb and woozy, my mind drifting away to sleep. In a panic, it occurred to me I hadn't locked the door.

Heart pounding, I tried to get up. I couldn't do it. My limbs had turned to liquid and no longer obeyed my commands. I couldn't blame them; I'd pushed them to the limit and then some. With one last nervous glance at the unlocked door, I fell into a very long and troubled sleep.

Unlocking

Over the next twenty hours I drifted in and out of consciousness, harried by nightmares, intense with vivid emotion and disturbing images. Sometimes I was being strangled and beaten. Sometimes watching, helpless, as Tequila's body dashed against rocks in a smear of blood. Most often I was falling. Either thrown high into a deadly nothingness or falling victim of snapping trees, uprooting bushes, crumbling rocks. Each time I awoke panicked and sweating, my throat dry and constricted, my head and heart pounding.

Belén had left a water jug and glass next to my bed. Stiff-limbed and unsteady, I spilled more than I drank. What I did drink, I sipped due to the sensation that shards of glass had been ground into the liquid as it passed through my swollen throat. Inevitably, though, my bladder filled and I had to get up to relieve myself. I struggled and staggered and winced my way to the bathroom, my injured leg ramrod-straight, the damaged knee seized-up, locked in place.

I peeled off my torn, mud-encrusted clothes and took a shower, washing my tender body as best I could. The hot water stung my cuts and scrapes and aggravated the hundreds of itchy bites covering every inch of my skin, but it was worth it; it felt better to be clean. I left my filthy clothes on the bathroom floor and hobbled back to the bed, where I collapsed again and pulled the covers up over my head.

The door clicked open and I awoke with a start. I couldn't have been sleeping long. My hair was still wet. I peeped over the covers, my belly tightening, sick with fear.

"Esta bien," soothed Carmen, stepping into the room, *It's okay.*

She carried a tray loaded up with small jars, cotton swabs, a steaming bowl of soup. Pablo was behind her, holding the door open. He smiled and nodded at me, then stepped aside and let another woman pass into the room behind his wife.

"Hello Jason," said the woman. "I'm Maria. I am Pablo's cousin. He called me here to find out what happened to you because I speak English and he does not. The policeman who brought you back to this place is coming soon, too. If you wish it, I will also explain everything you tell me to him."

"Thank you," I rasped, tears welling in my eyes. "Sorry," I gestured at my neck. "It's hard to speak."

"That's okay. There is no rush." She sat on my bed and took out a notebook with a floral-patterned cover. "Do you mind if I take notes on what you say? It will save you from repeating when the policeman arrives here."

I gave her the thumbs up.

As painful as it was to speak, it felt good to tell somebody what had happened to me, and some of the tension in my mind eased with the flow of words. While Maria scribbled her notes, Carmen set to work on my physical wounds with her salves and tinctures. Every now and then she would pause in her work, to listen as Maria translated my words or to feed me a spoonful of the broth.

When Carmen reached my busted knee, she frowned and interrupted Maria, who was in mid-flow, explaining my climb up the cliff-face to a bug-eyed Pablo. Maria heard what Carmen had to say and then turned to me. "We have to take you to a curandero who we know; a healer man. He will fix your knee. It is beyond Carmen's skill."

"The doctor said the x-ray machine at the hospital will be mended today," I offered. "Maybe we should go there first so we know exactly what's wrong with it."

Carmen gave me an indulgent smile and then shook her head. "Curandero," she repeated, while rubbing a pungent yellow tincture into the web of bruises on my leg. I noticed the mark of the vine that had saved my life was still there, imprinted in a spiral pattern from my ankle to the top of my thigh; irrefutable evidence of the impossible.

The police officer arrived and Maria relayed my story to him while Carmen finished her ministrations. He wrote down the translated statement, which I signed, and then he left.

It was the last time I saw him. There was never any investigation into my attack, and no effort whatsoever made to find my attackers. This complete disregard for what I'd been through, along with the willful irresponsibility of leaving two cold-blooded killers free to strike again, left me indignant and angry for the longest time.

After seeing I was as comfortable and well-fed as my condition allowed, Carmen drew my curtains and led Pablo and Maria away so I could rest. I must have fallen into a deep sleep because it seemed only moments later when the three of them returned. I hadn't heard them come in, nor had I heard them turn on my bedside lamp. It was dark outside, and Pablo informed me it was past nine at night.

"Jason, it is time to go," said Maria.

The two women left, and Pablo handed me a pair of clean shorts, a t-shirt, and a new pair of flip-flops. Once I was dressed, he helped me hobble outside to where a taxi was waiting for us at the curb. Maria and Carmen were already inside, squashed together in the front passenger seat.

Pablo dragged me across the back seat with one arm, while supporting my leg beneath the seized knee with his free hand. When I was comfortable propped up against the far door, he perched on the edge of the seat beside me and we set off across town. The journey was bumpy and painful, but mercifully short.

Our destination was a cinderblock shack with a rusted tin roof situated where the edge of town gave way to the jungle. A rectangle of blue light stretched from the doorway, illuminating a dusty yard full of people gathered in small groups, sitting on wobbly wooden benches, waiting. Pablo and the taxi driver carried me past the seated throng and into the shack, where an old indigenous man was tying a sling to support the bandaged arm of a plump, middle-aged woman.

Sitting on a shelf above the old man's left shoulder was a small boy of around five years old. He was barking instructions, or so it seemed, to either the man or the woman below.

I was placed on a white plastic lawn chair. Pablo, Maria and Carmen stood behind me, and we all waited in silence while the curandero finished with the woman. Just before leaving, the patient thanked the old man and handed over a five-dollar bill. He passed it up to the boy, who stashed it in a tin can by his side while making some kind of snide and amusing commentary. Everyone in the room smiled, including the woman with the damaged arm.

Once she departed, the curandero took a good long look at me. He glanced up at the boy and the two of them began giggling. In spite of the gulf in years between them, they looked almost identical when they laughed - all scrunched up eyes and bad teeth. The boy said something to me, which I didn't understand, and the old man laughed even harder. My friends snickered behind their hands.

"He asked where you want them to start," said Maria, sensing my unease. "Or if you just want his grandfather to find which bits of you are still working and start from there."

I chuckled as well. I guess I must have looked a hell of a sight.

Carmen spoke up, and the old man listened attentively. He rubbed the sparse, snowy stubble on his chin and bent over and felt my knee. After a few seconds, he gave his prognosis.

"It's dislocated, as of course you must know. But no bones are broken," explained Maria. "Don Chuncho is going to put it back to its place. It will take some time and it will hurt."

The little boy tossed something down to the Don Chuncho, who handed it to me. It was the end of an old leather belt. He waited patiently for me to figure out what I was supposed to do with it. I had a pretty good idea, but didn't want to go there. I couldn't really believe what was happening.

"He wants me to put this in my mouth, right?" I said, turning to Maria.

"Yes, of course. He doesn't want you to scream and make his other patients nervous." She gave a cheerful nod toward the open door. "It's harder to work on nervous people." She paused, and then added, "So it's probably better if you relax now, too."

"Relax? Are you serious? Doesn't he have any painkillers?"

"No. It's better if you don't numb the body when it wants to heal. Don Chuncho, he says painkillers make things worse."

"And a strip of leather to bite on makes things better?" I asked, incredulous, and more than a little afraid.

"¡Apóyate!" yelled the boy, setting off a fresh wave of laughter around the small shack.

"He says you need to support the pain," explained Maria. The curandero gave us a questioning raise of the eyebrows and Maria nodded. "Time to start," she said, giving me what I assumed she thought to be a reassuring pat on the shoulder.

I looked to Carmen and Pablo. They were the ones I knew and trusted. They were nodding in agreement and encouragement as well.

"Shit!" I hissed, and clamped the belt-end between my teeth.

What followed was the most excruciating half-hour of massage and manipulation, of cracking sinew and twanging tendons, that made me sweat and whimper and almost piss my pants. All the while, the little boy sneered at me from his perch, ordering me to *'support it!'*. The whole experience was so surreal, by the time the curandero had finished I wasn't sure if I wanted to weep or laugh with relief.

"You can take the leather out of your mouth now," said Maria.

I did, and the curandero took it from me. He inspected my teeth marks with an air of professional pride, then grinned and tossed it to the demon-monkey-child to chortle over. My knee was tingling and the skin and muscles surrounding it pulsated with heat. I figured he must have applied some sort of jungle heating salve, but didn't know for sure as I hadn't been brave enough to watch him work.

"Ahora!" he declared, with a clap of his hands. *Now!*

Maria leant forward. "He wants you to bend your knee."

I looked up at her, horrified. "I can't!"

"He says you can now. And you should."

So I did. And I could.

I was shocked. The joint was sore as hell, but it was supple; my lower leg glided back and forth without the slightest click or snag. I'd been given back complete mobility. Don Chuncho had fixed a dislocated kneecap and restored the full range of movement in less than half an hour. All I could think was how I wouldn't need to endure the lengthy process of surgery and recovery, months on crutches, physiotherapy. So much for the inferiority of primitive methods.

"Gracias," I whispered, taking his callused hand in both of mine and fixing him with what I hoped was an intense look of gratitude.

I didn't want to stop shaking his hand until he truly knew how thankful I was. The little imp vibrated with laughter up on his shelf, tossing jokes and insults down at me. The old man maintained eye contact with me, smiling politely, while at the same time explaining some things to my companions out the corner of his mouth.

"He is going to wrap your knee with some herbs for tonight. Tomorrow Carmen will remove them. He says you must walk as much as you can. Starting today."

The curandero applied a foul-smelling sludge and then wrapped my knee in a grayish bandage before pulling me up onto my feet. Pablo gave him a ten-dollar bill, and the old man insisted on giving us change before we were allowed to leave. After another gush of effusive thanks, I put a hand on Pablo's shoulder and limped out past the waiting people and into the taxi. The driver, who had been waiting for us all the while, smiled at me and took us back to the hostel.

Within a week, I could walk pretty well and was making daily laps of the hostel yard. Most of my cuts and bruises were healed, too. My neck and throat, however, were still sore and distended. It was difficult to eat anything more than soups, and my voice sounded odd; deeper and gruffer than before, with a strange loss of intonation. I'd never been the most expressive orator, but now my speech came out in a dull monotone no matter how much inflection I tried to put into it.

By far the most maddening and distracting of my injuries at this time though was my right eye. It was still closed up and it throbbed constantly, flashing with light every few seconds, making it almost impossible to sleep. The intense ocular flickering was often accompanied by a stabbing pain which would make the entire right side of my head pound for hours.

Aside from my physical state, I was finding it hard to let go of the trauma I'd been through. My days and nights were spent in a fog of miserable exhaustion, punctuated by panicked nightmares, debilitating fear and terrible rage.

It occurred to me most of my life had been spent in lower intensity versions of these states. I never had a clue what real happiness meant. I'd laugh and smile, but never a belly laugh or a truly contented smile. Fear and depression, resentment, anger, discontent - these were my emotions, and I could no longer fight or run or hide from them. My resistance was broken. I was so low, I might have taken my life if it weren't for Pablo and Carmen's kindness, or the sustaining hatred I felt for my attackers.

I'd tried to give Pablo the hundred bucks I'd had in my pocket when I was attacked, but he knew it was all I had left and insisted I keep it. They were housing me, feeding me, and nursing me back to health, and they didn't expect, or want, to be paid for any of it. When I assured Carmen I'd repay them once I could access my bank accounts again, she waved me away and explained, through Maria, that she felt responsible for me.

"You came to our country and met very bad men. You almost died because of their actions. Carmen doesn't want you to leave thinking all Ecuadorians are this way, hating gringos. We are good people, Jason, and we will take care of you."

"Thanks Maria," I said. "Tell Carmen I really don't think all Ecuadorians are bad. Most I've met are wonderful, welcoming and big hearted - especially you guys. I couldn't ask for more kindhearted friends."

I meant what I said. But that didn't stop the constant feelings of loathing and vengeance from eating me up. I hated the bastards who'd attacked me. They'd killed Tequila. A sweet and smart and defenseless puppy. And they'd nearly killed me. I wanted them to suffer and die screaming.

It didn't occur to me they might still feel the same way about me until the day I went out to the Internet café to write to Ravi and Fred

Miller. I'd lost my passport and all my belongings in the attack, and aside from the hundred bucks Pablo had refused, I had no money either. I needed help and felt Fred and Ravi were the ones who might provide it.

Fred lived in Quito, so I asked if he'd go to the American embassy on my behalf and see if they could begin the process of issuing me a new passport. I also wanted to tell him what had happened to me so he could warn other prospectors. The sooner word of my attack got out into the treasure hunting community the better.

I hoped Ravi might help me financially until my new passport arrived. Once I had official ID again I'd be able to access my local account, but until then I needed some cash to tide me over.

The glare from the computer screen set off renewed lightning strikes of pain across my eye, and it took a supreme effort of concentration to get the messages sent. Once it was done, I paid my bill and went outside. The day was overcast but not muggy. The fresh air felt good, and as it wasn't too bright out, I decided to go down to the river for a while.

Squinting against what light there was, I shuffled across the street to the pedestrian promenade by the river's edge. I found a bench and sat down and closed my eyes. Thankfully, it didn't take too long before the flashes receded to a faint misty kaleidoscope in the corner of my eye and the sharp pains subsided into dull throbbing. I listened to the river and the sounds of the people and the traffic.

"Jesus Christ on a bike, its Quasimodo!" said a familiar voice. "What you still doing here? I thought you'd be long gone by now."

Amy was standing somewhere off to my right. I turned my head so I could see her. "Hello Amy. Good to see you, too. Why did you think I'd be gone? I can barely move."

"Well, if someone beat the shit out of me and tossed me off a cliff, I might just take the hint I wasn't welcome anymore. In fact, you wouldn't see me for dust!"

"What? You think it was personal then?" Amy knew everything that happened out on the rivers. Of course she'd have heard what had happened to me. She might even know the guys who did it and where to find them.

"I don't think it was personal, per se. But I do know you gold robbers aren't the most popular folk around here. I may have mentioned it to you once or twice before."

"I think you might have, but more like one or two hundred times." This made her smile, and she walked around to sit beside me. "I'm not a gold *robber*, though," I insisted. "I work hard for what I get, and I don't pollute the rivers. Why would anyone want to hurt me?"

"Doesn't matter how hard you work or how spiffy you leave the waterways. Not to some folk, anyway. Granted, you're no Fred Miller, but you're still taking what doesn't belong to you. People get upset about that kind of thing here as much as anywhere."

The mention of Fred's name took me aback. It hadn't occurred to me Amy might know him. Like most treasure hunters, I guarded my information sources jealously. I mentally skimmed through the rest of her words and groaned inwardly. I didn't want to hear a lecture on how the land and everything in it really belonged to the local indigenous. I'd had enough of them and their bullshit double standards. But I did want to hear her opinion on Fred.

"Fred Miller?" I asked, trying to sound as nonchalant as possible with my rough, droning voice. "Who's he?"

Amy tilted her head, like a curious, or incredulous, spaniel. "The guy who sends jokers like you down here to Tena. Don't tell me you've never heard of Fred. His name is all over the treasure forums like a cheap suit. The bastard's been down here a few times himself, but only on the big operations. You know, the kind that destroy whole ecosystems but makes a few white guys really rich. The sort of shit that never seems to go out of fashion in the Amazon. Speaking of which, you never asked me why the gold you're taking really doesn't belong to you. Don't you want to hear the sordid truth of what you do? The reason you nearly got yourself killed?"

"Not really, Amy. I think I can figure it out for myself, thanks. I do want to know who attacked me and killed Tequila, though. Their names and where they live. You want to tell me that?"

Amy sighed and shook her head. "I'm sorry about Tequila, Jason. Seeing you without that little shite yipping at your ankles is sad to be sure. Worse than seeing you all banged up, truth be told. But it could have been a lot worse for you. And it still might be. You know they thought they'd killed you, right?"

I nodded. Tears filled my eyes, and I struggled to blink them back before she noticed.

"Look at you, Jason. For all your indignant anger, you're still a skinny gringo who couldn't fight his way out of a paper bag. Why do you want to go looking for trouble you can't handle?"

I opened my mouth to protest, but she cut me off.

"Listen, I like you. You're a good guy. But like most gringos who come down here, you haven't got a clue where you are and who you're dealing with. Those guys tried to kill you once already. Think about that. I'm sure they've heard you survived by now. And they know you could identify them if it came to it. I could try saying they all look alike to us whiteys and that you'd never recognize them in a million years, but why would they take the risk? You know what I'm getting at, don't you?"

"Of course. But I can't let them get away with what they did. Surely you can understand that?"

"They already got away with it, Jason. And so did you. Why not leave it at that? For all your bravado, if you do see each other again, you'll be the one to suffer, not them. I'd stake my kayak on it."

"Really? What makes you so sure?"

"Jesus boy, will you look at the state of you! You're half-blind, you wheeze like a broken accordion, and you walk with a wicked limp. A strong breeze would knock you over. Even fully fit they kicked your arse. Be realistic, your best bet is to get as far away from here as possible. They won't follow you. I can guarantee that. But if you stay in their territory, sooner or later they'll come for you. Do you really want to be reunited with your little doggy friend that much? Because that's how it'll end."

I knew she was right, but it didn't make it any easier to take.

"Just think about it," she added. "For my sake. But please don't take too long about it."

I nodded. "Thanks. I will."

I already was, and it was making my stomach churn and my skin prickle. Amy's words left me feeling exposed in every sense. What was I thinking leaving the hostel on my own? I wanted to be back in my room with the door firmly locked behind me.

As if reading my mind, Amy gave my arm a squeeze. "You want me to walk you back to your hostel? You nearly tripped over about ten times on your way across the street just now, and you might break your neck on the way back if you're not careful. Your reputation as a hard-as-nails survivor would be ruined if you did."

I smiled, grateful as much for her attempt at saving my pride as her offer to see me safely back. "Thanks. I'd love some company if you're going that way."

"Good lad. But don't expect me to hold your arm all the way now. People talk, you know."

*

I spent the next two weeks hiding out in my room. I was desperate to leave Tena, but Carmen insisted I take enough time to heal before attempting the journey to Quito. On the day she finally gave me permission to depart, the whole family drove me to the station and saw me safely onto the bus. Pablo and Belén waved, while Carmen and Maria crossed themselves and mimed karate chops at me; some sort of blessing, I imagined. I gave them a tearful smile in return, suspecting it would be the last time we'd see each other.

It was cold and dark and rainy when I arrived in Quito seven hours later. I had ninety dollars and the clothes I stood up in. Which unfortunately didn't include a coat. I jogged west out of the rundown terminal building to La Ronda - a narrow, cobbled lane of whitewashed colonial buildings that housed tourist restaurants, art galleries, museums, and Internet cafés.

I ducked into one of the cozy looking restaurants and ordered a warming glass of raw cane alcohol mixed with cinnamon and passion fruit, then went to check my email. Fred and Ravi had been in touch several times each. After my riverside chat with Amy, I hadn't been back to the Internet café in Tena and my lack of response had both men worried.

Fred told me in his first email he couldn't start the passport process on my behalf; I had to do it in person. He did offer to accompany me to the embassy, though, and insisted that I call him as soon as I arrived in Quito. Ravi agreed to lend me cash and offered to put me up in Vilcabamba, which, he wrote, was the perfect place for convalescing. I put on the headset hanging over the screen and called Fred from the computer. To my surprise, he answered the phone right away and then invited me to stay at his apartment.

His apartment was impressive. It had views across the city and an extensive library of treasure hunting, archaeology, geology and obscure reference books. Antique maps of Ecuador adorned the walls of his living room, and glass-fronted display cabinets showed off artifacts;

delicate gold masks, bone-handled obsidian daggers, pottery, yellowed manuscripts, stone figurines.

Fred himself was a tidy, compact man who I pegged to be somewhere in his early-sixties. He looked more like a scientist than an adventurer. Which I soon learned was appropriate. He was a geologist by trade and had initially come to South America working exploration contracts for multinational mining companies.

"I fell in love," he told me, as an elegant woman with coal black hair and warm dark eyes glided into the room. She said hello, then placed a bottle of Scotch whisky and two glasses on the coffee table between Fred and me. "Not only with my wife," he inclined his head toward the lady, who smiled back at him and left the room, "but with the entire country of Ecuador."

"I know how you feel," I said, while he poured our drinks. "In spite of what happened to me, I still love it here. The jungle, the wildlife, the people. Ecuador is pretty special."

Fred nodded and fixed his small watery eyes on me. "Glad to hear it. You're one brave man, Jason. I knew it'd take more than one little mishap to put you off your calling. I still think it incredible that you went so far out into the jungle alone." He raised his glass. "To fearlessness!"

We sipped our drinks, and I glanced around the room. Fred continued to study me. Something about him made me uncomfortable. He was polite and friendly enough, but his gaze had a predatory quality. Like a snake recoiling, readying itself to strike.

"You must have been out in the jungle by yourself a lot," I said, meeting his eyes. I nodded toward a framed newspaper clipping on the wall. "Not to mention crashing your helicopter there."

"*I* didn't crash it. But it was quite the experience. The pilot died, actually. The rest of us only just made it out. Not something I'd like to repeat, but not something that would stop my continued explorations either. As for going into the jungle alone, I don't believe I've ever been quite that courageous."

"Or that stupid," I suggested with a smile.

He laughed and poured another drink. "Don't be so hard on yourself. You're here now, that's all that matters. Now how are you feeling?"

I shrugged. "Not too bad. Throat still feels like sand paper and this voice is taking some getting used to. Worst thing is my eye, though.

It's annoying the hell out of me."

"How so?"

"Light flashes across it about every ten seconds. It's not too bad in the daytime but at night it's hard to ignore. Means I don't sleep much, and the tiredness makes it worse."

"Worse?"

"More painful. It's kind of like a migraine that waves in and out."

"Ouch. What do the doctors say?"

"I haven't seen one yet."

"Why not? Who took care of your other injuries? You said your knee had been dislocated, correct?"

I nodded. "A curandero fixed my knee. Another one, who happened to be the landlady at my hostel, treated everything else. She said my eye will right itself over time."

Fred's brow knitted together. "Brave indeed. But why wouldn't you go to a proper hospital? Have you gone native or something?"

"Not out of choice. I'm not complaining, though." I flexed my knee. "They did a great job, as you can see."

"I can," he admitted, begrudgingly. "But you can't. Perhaps we can go to my clinic after the embassy tomorrow and get your eye looked at properly."

"You have a clinic?"

He chuckled. "I mean the clinic where I go if I need something checked out. It's the best place in town. They'll fix it for you."

"Thanks Fred, but I think I'd better wait. I don't have any money for medical treatment right now."

"Oh, I see. Did the bastards take it all? Well, don't worry about that. I'll put it on my card and you can pay me back when you're on your feet again. Speaking of which, as you're heading south after here, perhaps you can check something out for me near Zamora. It could prove lucrative for us both. But you'll have to be discreet as there are a couple of bozos scouting the place already."

"Bozos?"

"Some aspiring hillbilly treasure hunters called, believe it or not, Dick and Willy."

I laughed, then something clicked. "Wait a minute, I think I might know those guys. Isn't Dick married to some Loja political guy's daughter? He and Willy have matching John Deere baseball caps, right? One's fat, one's thin. I didn't know they were into treasure hunting."

"Ah, you do know them. Well, that's embarrassing." Fred winked conspiratorially. "I was about to tell you I think they're retarded."

I laughed even harder.

"Seriously!" he continued. "They write me emails that make no sense whatsoever. I thought it might be some kind of code at first, but now I realize the pair of them are illiterate retards. Not the kind of guys I want to do business with; it's like pulling teeth, believe me. However, like a fool, I sent them to investigate a claim that's up for grabs on the edge of Shuar territory. You already know how much gold is around there, of course. Well, they were gibbering excited nonsense about the place in the beginning, but now they've gone very quiet all of a sudden. What would you make of that if you were me?"

"I don't know. Maybe that they're trying to cut you out of the deal."

"Trying to, I'm sure. Probably digging and washing without permission like the mentally-stunted hicks they are. The owner lives up here and wouldn't know if they were, you see. If you go take a look and see something going on then we'll know it's worth buying the claim. We could go in as partners. I'll purchase the land, get rid of Dumb and Dumber and arrange the machinery. You could secure the manpower and oversee the operation. What do you say?"

From what Amy had told me of his previous operations, I wasn't sure I wanted to be a part of anything Fred proposed. But I was staying in his house and drinking his whisky, and tomorrow I'd be going to his clinic to get my eye fixed, so I said, "Sure. I'll check it out."

"Excellent! Well, if you'll excuse me, I like to go to bed early. Inés has made your bed up in the spare room. Feel free to take a book to read. I'll see you in the morning."

He picked up the bottle and our glasses and left the room.

The next morning, we went to the embassy and I applied for a replacement passport. They explained the process would take a few weeks and that I'd have to pick it up in person. This would mean another trip to Quito.

"Not to worry," said Fred. "You can always stay at my place when you come back. I'm sure we'll have lots more to discuss by then."

Fred's clinic was a shining five-story hospital building on the opposite side of the city valley to his apartment. Inside, it was modern, chic and sterile - all pastel shades and gleaming steel. White-clad doctors and nurses bustled around, attending to the business of maintaining Quito's moneyed classes in the best of health.

"Fred, I appreciate your offer, but before we do anything, I'd like to know how much this is going to cost."

Fred patted my shoulder. "Of course. Take a seat while I check you in. I'll explain what's wrong and get us a quote."

Fred spoke with a pretty receptionist, who called in a glamorous doctor. The three of them had a conversation that involved lots of smiling and fake laughter. The doctor then gave me a little wave, before disappearing through a peach-colored swinging door.

Fred strolled back over, smiling broadly. "It'll cost the princely sum of a hundred dollars, and they're ready to start when you are."

I spent the next couple of hours going through various machines and scans and consultations. Eventually, I ended up back in front of the glamorous doctor, who prescribed eye drops, painkillers and dark glasses.

"She says it should calm down in time. If it doesn't we can always come back," said Fred.

I thanked the doctor and we went to pay and pick up the prescription. Fred gave them his credit card, and they gave us a slim pack of pills, saline eye drops, and a discreet caduceus-crested envelope containing the bill. When I opened the envelope, I saw we'd been charged over five hundred dollars.

"Fred, they ripped us off!"

"Did they?" He took the bill and looked it over. "Seems fine to me. It's itemized and everything is accounted for." He smiled at the cashier and led me away by the elbow.

"Yeah, but they said it would be a hundred bucks, not five hundred."

Fred shrugged. "They did a thorough job. That's what matters."

"All they did was tell me what I already knew."

Fred's brow creased into his now familiar frown. "Really, Jason, you did not *know*. All you had to go on was the conjecture of a couple of jungle witch doctors. Hardly a professional opinion."

"Maybe not, but they were still right and they didn't cost anywhere near five hundred bucks."

Fred clicked his tongue, indicating the conversation was over.

I stayed a few more days at Fred's place, looking through his books and talking with him about treasure hunting and ancient mysteries. He was an inspiring orator, passionate about his subject matter, infectious in his enthusiasm, but was very much the armchair adventurer.

I realized he was more spider than serpent. He sat in the middle of his web, waiting for the tug that meant one of his contacts had found something big enough for him to move on. The lone intrepid explorer image he cultivated online was a front to lure in people like me in to do his dirty work. The funny thing was I didn't respect him any less because of it, and by the time I left he had me all pumped up and ready to enter the wilds in search of riches once again.

*

"Good to have you back, Jason," said Ravi, when I stepped off the bus in Vilcabamba.

"Good to be back," I told him.

"Kitty's been asking after you," Ravi said as we climbed into his car. "Wondering when you were going to get here."

"Oh yeah?"

Ravi grinned. "Yeah. She started asking about a week or so before you emailed me. Did you write to her first? She seemed to know something was up."

Damn that girl was spooky. I shook my head. "She always knows when something's up. Goes with being a psychic, I guess."

Ravi nodded as if I'd just said the most reasonable thing in the world. This was one of the reasons I loved this quirky little town and its weird and wonderful inhabitants; the odd and eccentric were the norm. Not much could faze them.

Ravi came through with a loan and a place to stay. He even set up some work for me designing a small farm based on permaculture principles for a wealthy Canadian couple. The Vilcabamba magic didn't take long to work on my eye, either. By my second week in town the flashing and headaches had faded and all but disappeared.

Kitty said it was due to the reiki sessions she'd been performing on me. Ravi insisted it was the carrot-infused green juices Dana prepared us every morning. Curly put it down to the local microbrew beer he now stocked in his bar. It sure was good to be back in the valley.

After a couple of weeks working and settling back in, I wrote to Fred. I'd been earning well and wanted to send the money I owed for the hospital check-up. He wrote back telling me he wasn't in a rush for the money, but he would like to know what was happening to the claim he asked me to investigate in Zamora.

I took the next day off and caught a bus to Loja, then another to Zamora where I found a taxi to take me out to the area he'd described. There was nothing there. No heavy machinery, no dredging or sluicing, not even a pair of American bozos panning the river.

I thought about buying a gold pan to see for myself if the stretch of river was as rich as Fred suspected. But being at the edge of Shuar territory made me jittery. After surviving one attack, the last thing I wanted was to have my head cut off and shrunk by the notoriously hostile southern tribe.

On the bus back to Vilcabamba I met an acquaintance I hadn't seen in a while. Darcy was a Jehovah's Witness whom I'd met when he'd knocked the door of Ken Kerr's guesthouse one afternoon when I'd first arrived in Ecuador. As usual, he was dressed in the sect's uniform of white shirt, pressed gray slacks and guileless, shit-eating grin.

"Hey Darcy! Convert anyone today?" I asked, sitting down beside him.

"Maybe," he said. "But there's always room for one more in The Kingdom if you're interested?"

"Sorry, man. Not today."

After letting him hog the conversation for the first fifteen minutes, I had to change the subject from my soul's salvation, and began telling him what I'd been through since we'd last seen each other.

At some point I must have mentioned Fred and his opinion of Dick and Willy because the next day Willy turned up at my job site, demanding to know exactly what had been said about them. Apparently, he and Dick were Jehovah's Witnesses too. They were close with Darcy and his family.

"Look, I shouldn't have said anything," I admitted. "I'm sure Fred was only joking. You know what he's like. The guy has a sense of humor."

"Calling a colleague a retard ain't funny," he insisted. "Y'all ain't heard the end of this!" He then stomped off, muttering and cursing in a manner unbecoming a man of his religious convictions.

Unfortunately, he was right, I hadn't heard the end of it. The next day I received an email from Fred, rebuking me for lying and putting his reputation at risk. Willy turned up again in the evening, telling me pretty much the same thing Fred had said in his email. I took it from both of them, but inside I was fuming. I may have spoken out of turn, but I hadn't lied.

After this little drama, I wasn't surprised when Fred told me he was too busy to have me stay at his apartment when I returned to Quito a few weeks later for my passport. We did, however, arrange to meet so I could give him his money back. I figured I could apologize in person and make peace with him then.

I didn't get the chance. When I called, he told me he'd just walked in on some guys trying to rob his apartment. There had been a scuffle but the burglars had gotten away. Nobody was hurt and not much had been taken, but he and Inés were shaken up and awaiting the police.

"Shit, I'm sorry Fred. Is there anything I can do? If you want, I can come over and drop off the money. I won't stay, of course. Unless you want me to."

"No, no, Jason. We're fine. And as far as the money's concerned, I think you should keep it. You've been through a lot, and I'm doing very well financially at the moment - which is probably why we were robbed. Don't worry about paying me back. Call it a good faith gift from one treasure hunter to another."

I now felt even worse about breaking his confidence, and I vowed to myself if I ever did stumble on anything big, I'd make sure to bring Fred in on it.

I was in Quito for the whole weekend. I arrived Friday morning and was staying through to Monday when I was due to pick up my passport from the embassy. As much as Vilcabamba felt like home, being in Quito was always something I looked forward to, especially when I had a few dollars in my pocket.

I planned to take the cable car up the mountainside and hike to the Pichincha crater on Saturday, then wander the museums and galleries of the old town on Sunday. In between, I wanted to hit up the bars and enjoy the city's nightlife. I *had* been through a lot, and felt I deserved a bit of a blowout.

The hostel where I was staying was in a converted three-hundred year old building, with high ceilings and three foot thick exterior walls. It smelled musty inside, but the rooms were large, with hardwood floors, animal hide rugs, and solid furniture. The only concession to the modern age was a gaudy hot-pink rotating fan on the nightstand. Everything else could have been there since the place had been built in the seventeen-hundreds.

I was tired from the overnight bus journey so decided to take it easy on my first night. My plan was an early dinner followed by a couple of beers and reading in my room. I'd picked up a mountaineering book written back in the sixties from the library nook in the hostel's lounge. Full of antiquated descriptions of places I'd come to know well, I could think of no better reading material to get me in the mood for the next day's climb.

I went to sleep thinking of dirt tracks and rippling grasslands, craggy ridges and snowy peaks. I dreamed of an eagle flying in circles over pristine land, calling back and forth with a condor in a language I could almost understand, if I just concentrated enough. They were warning of a storm that arrived as a big, boxy-looking car, the exhaust deep and guttural, puffing smoke and sparks. Then an airplane roared in low overhead, setting off a car alarm and jolting me awake.

I got up and went to the bathroom. On the way back, I picked up my phone from the dressing table and checked the time. It was almost three. Too early to get up so I crawled back under the thick blankets and closed my eyes.

Sleep wouldn't come. I was too restless. My mind raced with random thoughts. Odd fragments from my childhood mixed with moments in the jungle - fishing, catching brightly colored bugs, gold glowing in a crevice beneath fast-flowing water, huge wonky blue butterflies.

I heard footsteps crossing the room above mine. Whoever it was had probably been disturbed by the car alarm as well. When they walked back and climbed into bed, the floorboards above my head creaked and the sound sent me back there in an instant. Where *there* was, I didn't know at first. All I knew was it was somewhere very real and terrifying.

In my mind's eye I see a chubby boy, his face lit up by a flashlight in the dark. Mikey, his name's Mikey. He's my friend, my best friend. Across a rundown rainy courtyard, I see two more faces. One is redheaded, with face full of freckles - Gary. The other has curly black hair - Chris. I know them; they're my friends as well.

Gary and Chris are worried about Mikey and me. They think we shouldn't be there. They're scared of what's looming, old and cold and gray, behind us. They're riding away on bicycles, Gary in a red plaid coat and Chris in a padded green ski-jacket. They leave Mikey and me behind. Alone.

Back in the darkness, something is watching. A door creaks. Someone shrieks. Mikey whimpers and yelps like a dog that's been kicked. Our flashlight beams jiggle as we run. Then they sputter to nothing. A feeling of foreboding turns to anger, to hatred, to revenge and death. Why does it want to kill me? What did I do? I'm so scared I cannot move.

A face in a high up window. Ancient, fierce, with long white hair - a demon. Why do I think that? Because it's coming for me and I can't escape. It crawls over me. Large and hateful and smelling of death, it breathes me in and spits me out. I run and scream, trying to reach my mom, my sister. I stumble on the stairs and all goes black.

These images and their accompanying emotions flooded through me with the force of a busted dam. They replayed like movie reels, over and over, in ever more detail, until by dawn, red eyed and weeping, I'd remembered it all.

It felt like a terrible nightmare, but I knew it wasn't. For one thing, I'd been awake the whole time. For another, it explained so much that had been missing all my life. The year of previously empty memories, the reason why I'd been sent to a kids' mental hospital, the lingering morbid allure of caves.

At the same time, so much was left unexplained. Why hadn't any of my family talked to me about it? Why wasn't I left with a phobia of caves instead of a fascination? Why didn't I hate Native Americans – because that's clearly what the man in the window had been - or spooky old buildings?

I had hundreds of questions and no way to answer them. Not true, I could speak with my sister and mom. But would they remember anything now? Maybe they'd recall the night terrors, because that's how it would have looked to them, but what about all the trips to *the ruin*? Could I find Mikey, or Chris, or Gary?

My head was spinning. I was just starting to get over the trauma of my attack, or rather I thought I was, but now I was back to square one. My hands were shaking and I was afraid to go outside. I was a mess of stifled emotions and jangling nerves.

I curled up on the big bed and stared at the sky through the window. Further small details from my formerly missing year in Arkansas flashed through my mind. Not just my trips to *the ruin* and their awful aftermath, but little things, fun stuff I'd done with my friends. I'd been happy and normal back then. In some ways that was the most shocking thing of all. There had once been a time when I didn't feel so miserable and dispossessed.

By the afternoon I was calm enough, and more importantly, hungry enough, to go out to a nearby restaurant. With food in my belly everything seemed a little better. I was able to think rationally enough to try make some sense of what had happened to me all those years ago.

It had clearly been some rite of passage thing that had gone wrong. But how it had gone wrong I couldn't quite work out. Had I been so terrified in *the ruin* that I'd subsequently hallucinated a scary old dude inhabiting the empty house next door? One who could teleport into my bedroom, rotten smell and all. As much as I tried to convince myself it must be something like that, deep down I wasn't buying it. It had all been far too real.

I forced myself to go to a local bar. I needed a distraction. I chose a place with a pool table and plenty of tourists and spent a few hours shooting pool and chatting with fellow Americans about nothing in particular. It did the trick. By the time I got back to my hostel at midnight, I had a nice buzz from the booze and the company and felt tired and woozy enough to be out before my head hit the pillow.

I dreamed of the jungle. I was walking to my camp in the Jatanyacu. Although I couldn't see him, I could sense Tequila was there with me. I was content, not frightened in the slightest. It was strange, I could remember the attack on my life but felt nothing whatsoever about it. No fear. No anger. Nothing. It was all part of the dream.

The foliage grew denser the further I walked, until it was hard to make out the path before me. There was a rumble deep in the ground and I felt the tremor in my legs reverberate up into my chest. The world shook, and thick, woody vines tore through the earth, shooting into the sky all around.

There were dozens of them, mottled and twisting, pushing ever upward, like Jack's magic beanstalk multiplied. I recognized them, their leaves, the shapes and textures; they were the same vines that had saved me when I'd been thrown over the cliff. They grew to immense sizes and formed an impossibly tall alleyway. I followed the path they made, trusting the vines knew where to lead me.

I didn't remember the details of the end of the dream, but I awoke feeling good; the vines hadn't let me down. It was at breakfast when I realized the sudden reappearance of my repressed memories was no longer playing on my mind. They'd been blocked out and now they were back. That was all. It didn't matter how accurate the memories were, or why they'd been unlocked after all this time, it just felt right to have them back where they belonged.

Falling to Fly

I picked up my passport the next day and went back to Vilcabamba. I paid Ravi back what I owed him and spent the next couple of months finishing the permaculture project for the Canadians and saving for my next expedition, without having any clear idea what it might be.

During this time, at least once a week, I had the same dream. Every time, the earth shook, and the vines ripped toward the sky, making a living passageway through the jungle. The vine dreams usually came when I was having dark thoughts about my attack or if I was obsessing about my unlocked memories. The day after the dreams I always felt better, calmer, more positive.

"What's on your mind?" said Kitty, one morning while we were having breakfast at the juice bar.

"You tell me."

We both grinned.

"You know me too well, Jason. But even so, conversations are much more enjoyable when at least two people are involved. Anything else is just mental masturbation."

I almost choked on my omelet. "Excuse me?"

"Excused." She looked at me over the rim of her coffee cup. "So, are you going to tell me what you're thinking?"

"If it'll bring your mind up out of the gutter." I told her about my unlocked memories and the recurring dreams. "I don't know if they're linked, but -"

"Of course they're linked!" interrupted Kitty. "Are you dense or something?"

"Probably," I admitted. "Care to enlighten me?"

"You're describing the ayahuasca vine. Sounds to me like the plant is calling you."

"A plant is calling me, huh? Okay. Any idea why?"

"Do you know what ayahuasca is?"

"I think so. I've heard it mentioned, anyway. It's hallucinogenic, right? Some kind of trippy jungle drug."

Kitty let out a melodramatic, exasperated sigh. "It's a plant medicine, not a drug. It's used to cure spiritual and emotional dis-ease. It's also very good at clearing out your bowels." She winked. "Pretty much anything you're holding on to gets purged one way or another."

"Sounds a bit too mystical and gross to me. Have you ever taken it?"

"Of course I have. I drink with a Shuar shaman on a regular basis."

This surprised me. "They let you drink their sacred brew?"

"This one does. His name is Manuel. He has retreats with foreigners pretty often. You want to come to the next one? You never know, it might be time for the vine to save your life again."

"Thanks, Kitty, but I don't need saving right now. You do know you're starting to sound a bit like Darcy, right?"

Kitty flicked her straw at me and a globule of green juice hit my cheek.

"Let me know when you grow some balls and I'll introduce you to Manuel. When Mama Ayahuasca calls - like she's calling you now, my friend - you'd do well to listen."

I didn't let on to Kitty, but she had piqued my interest. I started researching ayahuasca. At first, I found a lot of contradictory information. Even the meaning of the word wasn't agreed upon. Some translated ayahuasca as 'spirit vine' while others insisted it is properly called 'vine of the dead'. Needless to say, the latter moniker didn't fill me with much enthusiasm to partake in the brew.

What everyone could agree upon was that ayahuasca is a traditional and sacred medicine to the tribes of the amazon basin. It is prepared using the *Banisteriopsis caapi* vine and one or more other plants, most often the *chacruna* leaf. When the two are combined, they have a potent effect which is reputed to give access to the world of the soul and can help cure anything from parasites to mental illness to drug addiction.

That was the up side. The down side was severe, if temporary, psychological stress, something referred to as ego-death, and in some extreme cases, actual death. My only experience of drug taking was smoking a little pot now and then and taking ecstasy when back in college. I was a beer and meat and potatoes kind of guy. I didn't like to lose control. Ayahuasca sounded far too supernatural and dangerous for my taste.

I put the magic vine out of my mind and went back to gold prospecting. I had the same amount of good and bad luck as before, but even when things were going well, it brought me no real joy. I was slipping further back into the depression that had plagued me most of my life. Only now there was nothing to relieve it. No matter how much gold I pulled out of the land or rivers, I was never satisfied. In fact, I was growing to hate the sight of the stuff.

On one of my expeditions, I met a guy who also knew Fred Miller. What's more, he'd heard of me.

"I know you went through a lot," he said, echoing Fred's words from the last time we spoke. "But you shouldn't blame Fred for it. He didn't know you were going to get jumped out in the jungle. You should give back the money you took. He gave it to you in good faith. Ripping him off is pretty damn low if you don't mind me saying."

Even if I wasn't rendered speechless, there was nothing I could say to sway the man's opinion. Fred had a reputation. And, it appeared, so did I now. Neither of which was deserved.

The demonic face that had haunted my childhood was now let loose in my adult psyche. It merged with my attackers in nightmares and day terrors. The 'remember me' vine dreams continued, too. But by now I'd convinced myself not to trust them; I associated them with the return of my memories and the evil entity. Kitty told me the dreams and the unlocked memories were ayahuasca reaching out and working with my subconscious mind. If it were true, I couldn't help wondering what the plant had against me that it would wish to torture me so.

"Maybe you should just drink the bloody stuff," suggested Ravi, after we'd returned from another ill-fated expedition. "You never know, it might bring you out of your funk."

"You think it'll leave me alone if I do?"

Ravi shrugged. "Only one way to find out."

"Will you do it with me?"

"Hell no!"

I laughed. "I get it. Do as I say, not as I do, right?"

"Not at all. From what I hear, it can be really positive, but also hard. Like dying so you can be born again. Not in the cheesy Christian sense, but really empowering and transformative."

"So why don't you want to drink it?"

"No need. Don't forget I was knocking on death's doors when I arrived here. The experience changed me. Made me a better man. But I don't want to go through anything like it again."

I thought about the moment I was flung off the cliff. "I know what you mean," I said, and took a sip of my beer.

Ravi looked at me hard. "I don't think you do, Jason. But if you ever want to find out, go to Peru. The ayahuasca scene here is dark. I hear stories about Manuel, Kitty's guy. Not only does he take far too many people into his ceremonies and charge far too much money for the privilege, he's also been known to molest women while they're out of it. The guy's a sleaze. And that's not the worst of it. He has a reputation among his own people as a dark sorcerer, not a medicine man. A Shuar guy I know says he sucks out people's energies for his own nefarious purposes, then boasts about it back in his village."

"Shit, really? Does Kitty know about this stuff?"

"Probably. But you know how loyal she is. Anyway, I'm sure she has her own protection in place. She's a powerful woman herself in case you hadn't noticed."

"More powerful than me that's for sure. Between you and me, that's what I'm afraid of. I don't think I'm psychically strong enough to take anything like ayahuasca."

"Well, I think you are. And from what you've been telling me, it seems as though the vine agrees with me. Maybe that's why it snatched you from the jaws of the Jatanyacu."

There was nothing more to say. Ravi had said aloud what I was starting to believe already. I was scared as hell, but I was ready to give it a try. I finished my beer and stood.

"Thanks Ravi. As ever, your wise counsel is appreciated."

Ravi smiled. "You sarcastic sod! Seriously though, good luck, man. And make sure you find a good place with a proper shaman."

I nodded, but it was easier said than done. I looked online and asked around. There was a bewildering array of shamanic retreat centers in Peru. All of them were expensive and each of them boasted a more 'authentic shaman' than the rest.

I noticed a lot of the centers ran permaculture projects as well serving up ayahuasca and other medicines which cleansed and renewed body, mind and spirit. It seemed that ayahuasca and permaculture went hand in hand, which I took to be a good sign. I wrote to those centers that offered permaculture courses, asking if I'd be able to do a work trade in exchange for a ceremony. I also included a brief explanation of what had happened to me and why I felt called to drink the sacred brew.

Most wrote back saying they'd be happy for me to work in their gardens, but I'd still have to pay if I wished to take part in a ceremony. I crossed those places off my list.

It wasn't that I was being cheap. I wanted to make a fair exchange. Most importantly, I wanted to really get to know the people I'd be drinking with before I took the plunge. I didn't see myself as a typical tourist. I had time and energy and real expertise to give, and I wanted very little in return. All I asked was one ceremony in a safe and secure environment.

My time in Ecuador was drawing to a close and I still hadn't found a place in Peru, so I went back to the States to visit my family. I spent some time with my dad and Mary, and then went to see my sister, Kendra. I was desperate to find out what she could tell me about our time in Arkansas.

She didn't remember much. She did recall moving into a house a few months before we left for Texas. Apparently, she'd loved the place and hadn't wanted to leave, but I'd found it hard to settle there. "Not surprising, seeing as you still find it hard to settle anywhere," she added.

Next up, I went to call on my mother.

I didn't expect much from her. Years of medication to ease the pain of scoliosis and a score of botched surgeries had left her memory patchy at best. As it happened, though, she remembered more than Kendra. She remembered Mikey and my other friends. She chuckled as she recalled our Fatman and Ribbon nicknames and recounted stories about how much Mikey used to eat at our place before going home for dinner at his. This meant at least parts of my unlocked memories were real, if not all of them.

When I asked what had happened at the house, and why we'd moved away from Arkansas, her mood changed. She clammed up and denied any knowledge of me being unhappy there.

"You loved that place," she insisted. "We all did. I don't know why we moved, but I'm sure it had nothing to do with you." The look on her face convinced me otherwise. She definitely remembered something. She just wanted it left in the past. Which, soon enough, wasn't going to be an option for me.

When I checked my email later that night, I had a message from Peru. A guy called Yusuf was opening a new ayahuasca retreat center outside of Iquitos in the Peruvian Amazon, and he wanted me to set up their permaculture program in return for room and board and participation in ceremonies.

It was time for me to head back to South America again.

Jason Pednault   Matheu DeSilva

Demons Meeting

As soon as I received Yusuf's email, my mind turned to smuggling worms. The Amazon jungle, though famous for its lushness, has poor quality soil like most rainforests. To keep the nutrients from being washed away by the constant downpours, the plants and trees store them within themselves, leaving only a very thin layer of rotting mulch on the jungle floor. If a permaculture set up at the retreat center were to be a success I'd have to find a way to enrich the soil quickly and naturally. Which for me meant worm castings.

Yusuf told me there were no red wriggler worms available in Iquitos, but I wasn't too worried; I knew I could pick some up in Vilcabamba, and that the nearby land border between Ecuador and Peru was pretty relaxed. All things considered, I had the makings of a plan.

I decided to fly to Ecuador and take the bus - with the worms – down through the sleepy frontier town of Macara to Piura in Peru. Once there, I could travel overland to Pucallpa and take a boat downriver to Iquitos. I estimated the journey would take around ten days, and I wrote to tell Yusuf what I was doing and to give him my approximate arrival date.

In the end, things didn't work out as planned. I made it across the border without any problem, but when I arrived in Piura, the worms were in bad shape; they were writhing in a layer above the soil, starved of oxygen.

I'd perforated the lid of the five-gallon bucket where I kept them but it hadn't been enough. I hadn't anticipated the effect the violent changes in altitude we'd experienced winding through the mountains would have. Crossing Peru, and the Andes Mountains once again, by land would narrow the worms' chances of survival to nil. So I had no choice but to book a flight direct from Piura to Iquitos and hope the airline would allow me to bring three pounds of live worms as my carry-on.

They did.

Once I'd assured the customs officials my bucket contained common earthworms harvested in Peru, I was waved through as nothing more than an eccentric gringo with an odd worm fetish. This minor emergency meant I arrived in Iquitos a week early.

*

Iquitos wasn't at all what I expected. I had a vague idea it was the biggest city in the Peruvian Amazon but also knew there was no road access; to get there you need to travel by plane or by boat. I imagined this relative isolation would have limited its development, not to mention the amount of traffic and pollution. I couldn't have been more wrong.

From the minute I stepped off the plane, I felt sick. It felt like the whole city was assaulting my immune system with its smog and filth. My eyes were sore, especially my damaged right eye, my head ached, and my chest and limbs felt heavy. Breathing wasn't easy. Neither was thinking.

I took an airport taxi through the clogged streets to the hostel Yusuf had recommended. It was early afternoon, but I went straight to bed and spent the rest of the day and the long muggy night sweating through feverish dreams. The next day I wrote to Yusuf, telling him I'd arrived early and asking him to come get me. Thinking I'd be leaving for the jungle soon, I went out to explore Iquitos.

The screeching brakes, shouts, honking horns, squealing fan belts and revving engines made my head pound. As did the oppressive humidity and the stink of grease, garbage and piss. A three-wheeled tuk-tuk swerved toward me, the driver yelling, "Taxi! Taxi! Good price, amigo!" I shook my head, and he sped off, only to be replaced by another, then another, the same exchange repeated each time. Eventually I fixed my eyes straight ahead and walked on, ignoring their repeated calls.

The sidewalk was filled with beggars and hawkers. Barefoot indigenous children selling chewing gum and cigarettes slipped between the racks of sunglasses and gaudy jewelry being held up by their parents. Filthy hands and feet and pleading eyes were everywhere. Halfway down the street, a toothless man shoved a plastic soda bottle full of muddy liquid in my face.

"Ayahuasca!" he exclaimed. "Buy me and fly away, gringo!" His sales pitch ending with a demented laugh.

"Go away old man," said a young Peruvian with spiky hair and mirrored, aviator shades. The old man looked bewildered, and understandably so; the young guy had spoken in English - for my benefit no doubt. "You don't know what shit is in those bottles, amigo," he said, grabbing my arm and pushing a card into my hand. "But we got the real deal. Shipibo shaman. Real strong ayahuasca. Just what you're looking for, huh?"

"No, not really," I mumbled, trying to pull away.

"What you want then? A girl? Let Inti be your guide, my man, and I'll get you anything you want."

"No thanks, Inti. I appreciate it, but really I just need some food."

"Food? Perfect! I know the best places. Let me show -"

Taking my life into my hands for sanity's sake, I dashed out across the street. Ignoring the blasting horns and shouts of abuse from the drivers, I dodged between two tuk-tuks and a white panel van and made it to the other side where I ducked into a restaurant called *The Yellow Rose of Texas*.

As its name suggested, it was a gringo hangout. The coffee smelled great and the menu was decent. They even had an 'ayahuasca diet' section for folks about to head out to a retreat. I expected to be leaving town the next day but didn't think I'd be taking part in any ceremonies for a few weeks, so I sat down and ordered from the regular menu: eggs, bacon, toast and coffee.

Two hulking American guys were sat at the next table, their fat asses hanging over the seat edges. One had his back to me. His scalp was razor-shaved and sunburned, his bull neck creased and wrinkled. His companion was also burnt reddish from the sun. He had a triple chin that bobbed and quivered over the collar of his sweat stained safari shirt when he talked. His hair was cut in a military style with a blond tuft on top. Ketchup was spattered across his chins.

"Fifteen grand in three days and I didn't even have to touch a single board. Hell, I didn't even have to see any of the shit," said ketchup chins. "These guys here are so fucking stupid. They got no idea what timbers worth out in the real world. But they do work hard, I'll give them that. Cut it, hauled it out, packed it up ready for shipping, all for next to nothing. Man, I tell you, these Peruvian Indians are grade A idiots."

The bald guy chuckled. "You got buyers for it already, Ed?"

"Shit, yeah! You know me, Bob. I got buyers lined up for every stick they can cut down. Got the export permit guy in my pocket, too." He smirked and winked and snickered.

"So nobody cares where it comes from, eh? That's a goddamn

license to print money right there you lucky bastard!"

"Luck don't come into it." Ed smirked again. "You know that more than anyone. I see how your business works, too, you know."

The two of them chuckled in mutual admiration and resumed shoveling food down their throats. Bacon grease and syrup dribbled and joined the ketchup on Ed's chins. I had to turn away to stop myself from throwing up.

I couldn't believe what I was hearing. These guys were discussing illegal logging in a busy café in broad daylight, not to mention whatever seedy business Bald Bob was into. I had had this romantic notion of Iquitos being an ayahuasca-infused modern day spiritual mecca, but nothing could have been further from the truth. The reality was dirty and squalid. I was beginning to wonder what the hell I was doing there.

I checked my email when I got back to the hostel. Yusuf had been in touch, but he wasn't happy. He told me we had agreed the date for my arrival and he wasn't going to be able to pick me up until then. I was stuck in Iquitos for six more days.

I rubbed my eye. It was closing over and weeping; an infection, for sure. I tried to take a deep breath but my lungs only filled half way, and I coughed when I exhaled. Groaning, I trudged up the stairs to my room, locked the door behind me and went back to bed.

I spent the next few days in Iquitos shambling back and forth between *The Yellow Rose of Texas* and my hostel. I took care of the worms in their bucket, ate, tried to sleep, and most of all hoped and prayed I wouldn't die before I got to drink the ayahuasca. My eye was gunked up and itchy, and I had a chest infection and fever.

The day before Yusuf was due to collect me, a couple of tattoo artists arrived at the hostel. Hedgy (short for Hedwig) was a waiflike blonde from Seattle whose translucent skin was covered in flash art: roses, daggers, skulls, hearts. Her friend, Ben, who was tall and even thinner than Hedgy, had intricate scenes from *The Lord of the Rings* tattooed all over his torso and up and down his arms. The two of them were on the strict ayahuasca diet and accompanied me to the restaurant for dinner.

"What's up with your eye, man?" asked Ben, once we were seated and had ordered our food.

"Looks like pink eye. Is it pink eye? Are you contagious?" Hedgy moved her chair back.

"I don't know. Could be."

"Actually, never mind the eye, what's up with your voice?"

"Hedgy! Give the guy a break. Maybe it's always been like that." Ben turned to me. "Has it always been like that?"

I smiled for the first time since arriving in Iquitos. These two were quite the double act. "No, it's a recent thing," I said. "And a long story."

Hedgy leaned forward. "Have you got throat cancer? Is that why you're here in Iquitos?" Ben was about to open his mouth, but I held up my hand, letting him know it was okay. Hedgy noticed and shot him a look. "What Ben? Plenty of people take aya to cure cancer. Maybe Jason's one of them."

I laughed and coughed, then gasped and cleared my throat. "I don't have cancer. At least I don't think I do. But this week I've felt bad enough to imagine I have all kinds of things."

"You mean you were fine until you got to Iquitos?" Ben passed his hand across his face, touching his right eye then his neck. "The eye, the voice, the cough, the wheezing?"

"Everything but the voice."

"And that's the long story," said Hedgy, eyes lighting up. "You should tell us tonight as we're going out to the jungle tomorrow. Our friends Tori and Yusuf are picking us up and taking us to their new retreat center."

"You know Yusuf?"

"No, we don't," Ben answered. "Not yet. But Hedgy tattooed an om symbol on Tori's ankle just before she and Yusuf got together."

"And now she wants the aya vine tattooed up her leg," added Hedgy.

I thought of the twisting bruise the vine had given me the year before. "I had one of those once," I said.

Ben and Hedgy frowned as one. The effect was astonishing, making them look like twin brother and sister. Ben was first to articulate what was on both their minds:

"Eh?"

"Part of the long story. I'll tell you when we get to the center. I'm going too." Both their mouths dropped open like a pair of matching ventriloquist's dummies. "Are you two related?" I asked.

They shook their heads, looking more alike than ever. "Just work together," said Hedgy. "You doing the ceremony with us?" She looked at what was left of my steak. "You do know about the dieta, right?"

"I do, but I won't be drinking for a while yet. I'm going to be setting up the permaculture side of things for them."

At this they both nodded. Then Hedgy broke into a wide grin, without letting Ben know it was time to change their facial expression. "I bet you do end up drinking with us. How could you not?"

Tori and Yusuf arrived the next day in a shiny new minivan. Tori hopped out onto the sidewalk and beamed at us all. She was in her mid-thirties, bottle-blonde and crisscrossed with beads and flowing scarves; she looked like a hippy Christmas tree. Her white top billowed out, and Yusuf, who'd just walked around from the driver's door, placed his hand on the bulging fabric and guided it back to her skin while at the same time deftly pushing her to one side.

"Jason! Good to meet you at last," he said, extending his hand to me. "I'm sorry it's taken a while. We've been rushed off our feet getting things ready for the ceremonies this weekend." He glanced at the bucket by my feet and raised his eyebrows. "I'm guessing those are the wriggly worms that caused all the trouble." He smiled like a movie star, all perfect teeth and tan skin. He reminded me of a young Antonio Banderas. I couldn't help but smile back.

"They'll be worth it, you'll see," I said.

"Someone has an ego," Tori sang, with a head wobble she no doubt picked up in an ashram in India.

Yusuf wasn't fazed by his girlfriend's peculiar rudeness. He kept smiling and shaking hands; the man was a consummate schmoozer. "Let's get going, shall we?" he said, opening the van's side door and stepping aside for us to climb in.

"Where are you from, Yusuf?" asked Hedgy, once we were on our way. "I can't place your accent."

"México. But I grew up all over. Mostly in the States though."

"How did a Mexican-American end up opening an ayahuasca center in Peru?" said Ben.

"Long story, my friend."

"There's a lot of that about," said Hedgy, nudging me in the ribs.

"Actually, it's a calling," said Tori. "Mama Aya told him to create this center while he was in ceremony."

"Wow, seriously? That's some great financial advice from a plant," said Ben. "Can't wait to hear what she's going to tell me!"

Tori rolled her eyes at Yusuf. "Be careful what you wish for, Benjamin. Ayahuasca will humble that proud western sense of humor.

She'll make you realize how truly insignificant you and your jokes are."

"Doesn't seem fair considering how much it costs," Ben retorted.

"The price is an offering. The more you give, the more you'll receive."

"Guaranteed bang for your buck, eh? Well that's a relief."

I tried to stifle a laugh and ended up in a coughing fit and didn't hear Tori's response. Once I'd recovered, I realized everybody was staring at me. Even Yusuf, who utilized the rear-view mirror to do so. I mumbled an apology and asked Tori what she'd been saying.

Tori sighed and gazed out through the windscreen. "Are you sick, Jason? You shouldn't be going into the jungle if you're sick. Tell him, Yusuf. The jungle is no place for sick people. It's harsh and unforgiving. Maybe we should drop you back at the hostel until you get better."

"No way." I said. "It's the city that's made me sick. I know the jungle. I'll be fine there. I can handle it."

"That's your ego talking. You have a huge ego problem, Jason. You know that?"

I was taken aback. I'd always thought lack of confidence was more my issue. "Trust me," I said. "I really don't think I do."

"Well, that settles it right there then. Your ego is so big you don't even see it anymore. Isn't that right, Yusuf?"

Yusuf shrugged. "We all have our problems, honey. This is why we are here. To learn from the wise Mother and heal ourselves."

"Which is also why we shouldn't take Jason back to the hostel," insisted Hedgy. "He's sick. And we're on our way to a healing retreat. What could be more appropriate?"

Tori slouched into her seat and pouted her perfect lips. "A little less ego would be nice."

Yusuf patted her thigh. "Mama Ayahuasca will take care of it, honey. Don't worry."

"I know, baby. But ever since we let go of our own egos, I find it harder and harder to have to deal with all these egomaniacs."

I glanced at Hedgy and Ben. They were wearing identical expressions of amused bewilderment. Right then the three of us probably looked like triplets.

We picked up a few more people then headed out of town on a highway which had been built to service logging and gold mining sites. After an hour or so we pulled onto a dirt road and parked.

"We have to hike in from here," said Yusuf. "It's too muddy for the van."

"Is this a fire road?" asked Ben.

"No, not in the rainforest. No need. It's too wet out here for that. This is an old logging road. That's why it's so rutted. Heavy trucks," Yusuf explained. "This one isn't used so much anymore. Now it's more of a three-mile long private driveway for us."

I thought of Ed - the illegal lumber dealer from the restaurant - and my heart sank. Was there nothing sacred in this so-called sacred place?

Once we began walking, however, my mood lifted. Well-preserved old growth forest towered on both sides of us, alive with the chatter of monkeys, the flutter of tropical birds. After the hellish week I'd spent in Iquitos breathing noxious fumes and surrounded by obnoxious people, this was exactly what I needed. As rundown as I felt, the three-mile hike passed in no time at all. The raw nature was energizing me.

"There's going to be a sign here," said Yusuf, pointing up as he veered left through a wide gap in the trees. "But it is not ready yet. Everything else is good though. As you'll soon see."

He was right. The center was beautiful. A stepping-stone path made of cut logs led down from the soon-to-be parking lot into a wide natural bowl. Streams trickled between gigantic trees and the low shaggy foliage of what Yusuf told us were native medicinal plants. A large thatched roundhouse with low walls of beaten bamboo and green fly-screens sat up on stilts on one side of the vast bowl, with miniature versions of the same construction scattered elsewhere throughout the grounds.

Yusuf nodded toward the smaller buildings. "Your accommodation: traditional, jungle-style tambos. Each has its own features and character; dense jungle, big trees, streams and bridges, some high up, some deep down. We'll assign them in accordance to your individual energies. The large tambo is the maloka, our sacred ceremonial space. Beyond is the common area, where we'll take our meals together and prepare the ayahuasca. There are massage and healing rooms over the ridge. Take the path to the left. Don't go right, or you'll end up either in the kitchens or the workers' quarters."

There was some polite chuckling and then Tori took over. "Due to the amount of people we have here this weekend, some of you will be sharing tambos. But don't worry, we'll be matching roommates energetically, too."

"Sounds exhausting," quipped Ben.

Tori scowled at him. "For example, Ben and Jason, who are both sick and need to do some serious ego work, will be in our dense jungle tambo."

Ben shrugged and I grinned. Dense jungle sounded great to me.

I was too spellbound by the magic of the place to pay much attention to the rest of the orientation. Yusuf and his crew had created a stunning jungle paradise. Not that the jungle needs much help in that regard, as far as I was concerned, but the sensitive way in which the buildings and paths blended into the environment was inspirational. I was excited to think this would be my home for the foreseeable future.

*

"What do you make of these digs?" asked Ben once we'd arrived in our tambo.

"Amazing. Reminds me of a place I went to once in Belize. It's like there's no barrier between us and the jungle outside."

"That's because there's not. And I'm not sure that's a good thing. Aren't there wild animals out here? Like jaguars and snakes?"

I laughed. "I wouldn't worry too much. We probably won't see them even if they are there. Animals are a lot less dangerous than people, you know, and they avoid us when they can. That being said, we should keep the screen door closed to keep the mosquitos out."

"So, you're really not bothered about the flimsy walls, huh? I bet you don't care about the hot, humid air, damp sheets, mildew and lack of Wi-Fi, either."

"Hell no. I feel a lot more comfortable here than in any city, that's for sure."

Ben shook his head. "I thought you and I were supposed to be the same, 'energetically'. I'm a city boy and you're a nature freak. We couldn't be more different! I think Tori only put us together so I'd catch your conjunctivitis."

"Duh, of course she did. That, and she wanted our raging egos as far from her and Yusuf's palace as possible."

Ben chuckled. "You saw the luxury tambo, then? Hard to miss the hardwood shutters, Tibetan prayer flags, Ganesh posters and faux-

Persian rugs, wasn't it? Not to mention the flashy sound system, MacBooks, and the fact it's the only one of these half-finished sheds on ground high enough to catch a breeze."

"Yeah, they do have a pretty nice set up. It was probably a reward they gave to themselves for letting their egos go. Hey Ben, you mind if I ask you a question? If you don't like the jungle, and you think they're charging too much money for what they're giving, what are you doing here?"

"You mean apart from trying to lose my nasty western sense of humor? It's simple really: I love drugs. All kinds. You name it, I've smoked it, snorted it, popped it; even shot it up. I'm also a part-time dealer back in Seattle, and therefore have a professional curiosity to know what all the fuss is about."

"There's a fuss about ayahuasca in Seattle? I guess I'm a bit out of the loop of what's going on in the States these days."

Ben laughed. "Not among your sort. But it's getting to be a big thing in certain circles. People pay thousands to come out here and go tripping. Most say it's a spiritual thing, medicine, and all that crap. But they really just want to trip-out, I'm sure of it. I know my clientele, after all.

"What I want to find out is what's so special about this jungle juice that dealers like Yusuf can get away with charging bank for a glass or two along with some basic accommodation in the woods. I mean, it's not like they even have to smuggle anything into the States! Everyone comes here. To the source. Which reminds me, there was a dude with no teeth selling big Coke bottles full of the stuff for five bucks outside our hostel. If it's any good, I'm going to try take some back with me."

"You don't know if he was selling the real thing though. Pardon the pun. And shouldn't it be taken in a ceremony?"

"Why? The whole ayahuasca retreat thing is new here you know. It's just for gringos and about as authentic as Disneyland. I bet the local guys don't force themselves to eat shit food for two weeks before drinking aya either. My Spanish is pretty good. I'm going to ask them about it as soon as I get the chance. Anyway, enough of me and my ranting. Why are *you* here? And don't tell me it's a long story or I'll strangle you in your sleep tonight."

"Okay then I won't. I'm here because I love the jungle."

"Ha! Good one. You're a funny little shit, you know that?"

"Not until now. I'm going for a walk. You want to come?"

"Nah, you go explore. I'm going to stay here and smoke a joint."

I wandered out past the area of log stepping-stones and found a path leading further into the jungle. I wandered between the ancient moss-furred trees, watching lizards scuttle and butterflies lope through the air.

The canopy high above teemed with invisible life. I caught an occasional glimpse of a monkey or the flash of bright feathers, but for the most part all I could do was hear it. So, I stopped, closed my eyes, and listened. After a few minutes of the squawks, chirps and rustling, I heard something else: it was the sound of somebody singing.

I opened my eyes and immediately saw the singer way off through the trees. It was a white guy with a wispy ponytail and a bright orange t-shirt; hard to miss in the jungle. He was bent over, collecting leaves and placing them in a wicker basket at his side. His song was a rhythmic, otherworldly chant, unlike anything I'd heard before.

I moved in closer so I could hear better. I noticed he was picking the leaves one-handed. His other hand rested lightly against his throat. When the song ended, I clapped. The singer ignored me and continued with his work as if I wasn't there.

I was about to leave when he turned around and looked at me. His eyes were unnerving, pale green and intense. I had a feeling of déjà vu. He reminded me of someone from my childhood, but I couldn't place who.

"Yes?" he asked, staring intently at my face.

"Sorry, I heard you singing and thought I'd come over."

He nodded, keeping his disconcerting green gaze fixed on me. "My friend, there is something seriously wrong with you. You know that, right?" His speaking voice was odd, almost robotic in its precision. "You need a full reset for your health. I think sanango." He tilted his head and looked me up and down. "There is something else. Maybe related. Must be. All things are, after all." He held out a pale, delicate hand. "I'm Alf."

"Elf?" I said, mishearing him. I was worried I might crush his fingers when we shook hands, but his grip was surprisingly firm. He was stronger than he looked. "My roommate would love you. He's got elves and dwarves and dragons all over his body."

He smiled. "Sounds painful. But I didn't say Elf, I said Alf. Alfie. Alfred."

"Oh, okay. Sorry about that."

"Not a problem. Usually it's me who gets names wrong. I'm stone deaf, you see, and it's hard to read a mumble."

"Oh, right. I'm sorry. I didn't realize. You don't sound deaf. I mean, you were singing. How can you sing if you can't hear?"

Alf raised his fingers to his throat. "Vibration. Thanks for the compliment. I work hard at my speech, as well as my icaros; the native song you heard me singing just now. Now please stop apologizing. You're starting to sound British."

I laughed. "How do you know I'm not?"

"I can read your accent. You're American through and through, just like me."

I nodded, impressed. "I'm Jason, by the way. I work here."

Alf raised a feathery blond eyebrow. "So do I. I'm surprised we've not met before."

"Yeah, well, I only arrived today."

"Ah, right, you're the permaculture guy. Good to meet you. I work with Don Sharaco, the shaman. I meant what I said before. You will need some serious attention before you can be of any real use around here."

"I'm a bit run down, that's all. I was fine before arriving in Iquitos."

Alf shook his head. "I doubt it. Let's go see Yusuf and see if we can get you straightened out."

Alf led me to Yusuf's tambo and explained what he thought I needed to our boss. Yusuf listened, then turned to me.

"I don't really have time for this crap, Jason. Do you really think you need a ceremony and cleanse before you can start work?"

"No, not at all. I'm ready to start now. Tell me what you need me to do and I'll get on it. I can do the cleanse when things calm down a bit."

Yusuf exhaled. "I don't really have time for that either. Can't you use your initiative?"

"Sure. Happy to. Point me in the direction of the tool shed and I'll get started."

Throughout our exchange, Alf had been reading our lips, moving his head back and forth as if watching tennis. Now he was vehemently shaking his head. "Jason needs a cleanse as soon as we can do it. He can't be here like this. It's bad for the vibe of the ceremonies."

"There's nothing wrong with *my* ears, Alf," Yusuf snapped. "I heard you the first time. Don't worry. He'll get cleansed. Let me chat with

Geraint and see if we can fast track him through an aya ceremony so you can get started on him." He turned back to me. "But don't go running away once you've had your ceremony, though. We have a lot to do here. Now, if that's all, guys, I have work to get back to."

Yusuf disappeared back inside his tambo.

"Who's Geraint?" I asked Alf.

"The other boss. Moneyman from Europe," Alf turned away toward the ridge. "Let me show you what tools we have. Although I really don't think you should be doing anything until you're mended."

Mended seemed an odd choice of word. But Alf knew exactly what he was talking about.

*

For the rest of the day and the following morning, I mixed soil and prepared seeds for germination. I'd brought moringa, corn, squash, watermelon, cucumber, tomatoes; everything I thought would do well in the jungle.

What I hadn't brought were my glasses, and with the infection now spreading into both eyes making it too irritating to wear my contacts, I was effectively blinded. So, I huffed and puffed to and fro, making boxes for the seedlings by feel. It was monotonous work, and time consuming in my impaired state, but it felt good to be doing something of use.

After lunch, Alf invited me to help prepare the ayahuasca for drinking the following night. All the participants in the ceremony had to have a hand in the preparation, so it was quite a social, if somewhat somber and respectful, process.

I was building up the fire when Alf pulled me aside and applied a crude salve to my eyes with his thumbs. After a few stinging seconds, my eyes stopped itching and my vision cleared a little.

"What did you just do?" I asked, still blinking.

"Saved all our lives. A blind man really shouldn't be allowed to tend a fire, you know. It's kind of dangerous."

"That's right," said Hedgy, who I could now see was sat across the fire from me. "You should always listen to the deaf guy. He sees things better than anyone." She tapped Alf's arm and gave him the thumbs up. "Great job on the makeup, Alf. Mud brown mascara really brings out the red in his eyes."

Alf asked if I wanted to help prepare the vine so I followed him to where the long pieces of woody ayahuasca were laid out on the grass. Ben was there, chopping a thick length into smaller pieces with a machete. Another guy, Phil, was pounding the gnarly knots out with a mallet. Seeing the vine that had saved my life being butchered and beaten for the brew made me feel somewhat uneasy.

"What's wrong?" Alf asked, reading the look on my face. "We have to do this or it won't fit in the pot."

"I know. Seeing the vine just takes me back though."

"Back where?"

"You don't know why I'm here? Yusuf never said anything?"

Alf shook his head.

Ben and Phil had stopped working so they could listen in on our conversation.

"Now might be a good time for a long story, Jason," suggested Ben. "After all, we've got a lot of work to do and it'll help pass the time."

So, while we worked I told them how the vine had saved my life back in Ecuador, and how it kept appearing in my dreams ever since my repressed childhood memories had come back. By the time I'd finished, we were ready to combine the ingredients in the water and start the brewing.

Before we did, Alf fixed me with his discomforting stare. "She's not been torturing you with these dreams and memories. Ayahuasca, I mean. Your psychic friend in Ecuador called it right: ayahuasca's been working with you from afar. You have to drink tomorrow so we can begin your healing. I'll square it with Yusuf."

I nodded and thanked him, though deep down I was still unsure if I wanted to take ayahuasca so soon. I'd planned to wait until I knew everybody at the center and felt secure they'd be there for me if something went wrong.

I liked and trusted Alf, but we'd only just met and I didn't know what kind of role he'd have in the ceremony; he might be too busy to be able to watch out for me. Besides, I was too sick to be taking a mind-altering substance. Aside from my infected eyes, my bones ached and my lungs felt like leaky balloons. It made more sense that I heal before drinking. Not the other way around.

*

The ayahuasca mix needed all night to brew. We each took turns stirring the giant pot throughout the evening and then the center workers took over for the nighttime vigil. The participants, including me, were told we needed to rest before the following night's ceremony.

Alf gave me a small clay pot containing the balm he'd put on my eyes. I smeared some over my eyelids before bed and again when I woke up in the morning. My eyes were still too gritty and sensitive to put in my contact lenses, but they did feel better, and I hoped my other symptoms might also abate once the eye infection cleared up. *Then* I'd be ready for a ceremony.

I went to see Yusuf, to make a plan for the vegetable gardens, and to tell him I'd rather defer drinking ayahuasca until I felt better. He wasn't in his tambo, but Tori was. She told me Yusuf was in Iquitos, picking up some more people who wanted to do the ceremony.

"Doesn't everybody taking part have to have had a hand in the preparation?" I asked.

"Wow, Jason. You think you're an expert on how things should be done here already? Oh. My. Goddess. You really need to check your ego, my friend."

"I don't think I'm an expert. It's just what Alf said yesterday."

"Alf is not the shaman. He's a trainee-shaman, one who's getting a bit too pushy too since he started hanging out with you."

"What do you mean? I only met him yesterday."

"I know. And already you've got him telling Yusuf to let you drink before you've done any work here."

"That was his idea, not mine. I'd prefer to wait anyway. It's one of the things I came to talk to Yusuf about actually. The other is to discuss where we're going to put the vegetables and the trees I have germinating. I don't suppose Yusuf spoke to you about that has he? I could do with a bit of guidance."

"Goddess help me, your energy is too intense! I don't know where you should put the trees. Don't they just grow out the ground where they're supposed to be? I thought you were supposed to be the tree expert. Isn't that why you're here? And now you're telling us you've decided to back out of the ceremony when it's too late to find someone to fill your place. That's really selfish, Jason. Your ego really is too much."

"So you keep saying. Maybe that's why I'm here - so I can lose this terrible ego problem you think I have. Have you ever thought of that?"

"I have actually. Yusuf spotted it on the first day. You seem to be here for yourself and what you can get. Not for the greater good of the center."

"What? Okay, Tori, whatever. I'm going to go back to work now. When Yusuf gets back can you ask him to come find me please?"

Tori shook her mane of bleached straw hair as if auditioning for a shampoo commercial. "I can ask, but don't expect him to do it. He's going to be far too busy to talk to *you* today."

Now it was my turn to shake my head. Furious, I stormed off to my soil mixing area. Nothing calms me more than getting my hands into some dirt and working with plants. Still, it took the best part of an hour for the soothing effect to take hold after my encounter with Tori. But take hold it did, and I spent a blissful few hours mulch-making and transplanting seedlings in the cool shade of my work space.

Not long before lunchtime, I felt a sudden stabbing sensation in the right side of my upper abdomen. I winced and waited for the pain to subside. It didn't though. Instead, it grew more intense. I doubled over and toppled to the ground. The pain was centered in front of my right kidney and all I could think was something serious must have ruptured inside me. Either that or Tori had fashioned a voodoo doll with my likeness and was ramming a sharp needle in and out of it.

I caught my breath and screamed for help. Nobody came. The pain was unbearable, and I twisted and writhed and sobbed, wondering what the hell was happening. My vision grew grainy; I was close to blacking out. My sobs transmuted into long, low moans. I shivered, feeling cold. So cold. Cold, in the jungle. It didn't make sense. Nothing made sense, apart from the fact I was sure I was dying.

I screwed my eyes shut and panted, trying to breathe through the pain. But it was impossible. I heaved and retched, eyes bulging wide. Startled, I jerked back. Hedgy's face was inches from mine, her brows knit together in concern.

"Are you okay?" she asked.

"What do you think?" I hissed back. Her frown deepened and I softened my tone as much as one can through gritted teeth. "My stomach. Or something else. I don't know. It hurts so much. Please, get me some help, Hedgy. Please!"

She jutted her chin then disappeared from view. I resumed groaning and listing in the dirt until she came back with Alf and Ben and some of the workers. They carried me to the maloka and I was placed on a hastily prepared bed of cushions. Alf muttered some instructions to one of the workers, who ran off out the doorway. He then told everyone else to leave.

Once they were gone, he narrowed his eyes and wiped my forehead with a small white handkerchief. It came away grimy. Sweat and dirt. "You can scream if you think it will help," he said. "Everybody else has gone and you know I can't hear you so it won't bother me."

"Thanks. But it's starting to ease off now. I don't think -"

The pain came scything back in with renewed intensity and I did scream. In fact, I howled and lurched and screeched and juddered across the floor. Alf gripped my hand and I squeezed back hard. He grimaced but didn't pull away.

"Don Sharaco is coming. You'll be fine. He'll know what to do," he reassured me once I regained control of myself.

"Why?" I panted. "He's the shaman. I need a doctor. Not a shaman! Please, Alf. I think I'm dying. Seriously. Get someone else."

"You won't die, Jason. Not yet, anyway. You say shaman, we say curandero; which means healer. Stop worrying and save your strength."

I thought of the old curandero and his mischievous demon grandchild in Tena. He'd been able to work wonders, painful wonders, but wonders nonetheless. Maybe Don Sharaco had had similar training. I nodded and set my jaw as another thunderbolt of pain exploded through my belly.

The next two hours were the most agonizing I'd ever experienced. A part of me hoped I was dying. At least dying would mean an end to my torment. Alf stayed with me throughout, whispering encouragement and wetting my lips and brow with fragrant water from a clay pot one of the kitchen workers had brought in.

Don Sharaco appeared and came straight to me. He was a squat, gray-haired, solid man. His clothing was traditional Shipibo, with bands of brightly colored patterns on a background of white. Even in my distracted state, I could feel his immense presence, and it gave me some confidence things might turn out alright.

With a nod to Alf - who withdrew at once - Don Sharaco tied a headband around his forehead and held his arms out over me. He muttered and chanted something with his eyes closed, and then turned to receive some things from the wizened old woman who'd arrived with him.

The woman stepped back and stood with Alf, while Don Sharaco spun back around and stretched his arms out again. In one hand he now held a bunch of leaves, in the other, a loose-rolled cigar that wafted smoke and flaked ash over his forearm. He shook the leaves over me like a maraca while taking puffs on his cigar and blowing the smoke in my face.

I wanted to cry. The pain in my belly was growing worse again and this so-called curandero was doing nothing but making me feel queasy with his stinky cigar and shaky leaf act. Just when I thought things couldn't get any worse, he took a swig of some rank liquid and began spit-spraying it all over me. It smelled like an alcoholic's ashtray and made me gag.

I closed my eyes and trembled against the stinging in my gut and the wild anger rising in my blood. Don Sharaco kept shaking his leaves. He began singing an icaro, similar to the one I'd heard Alf sing in the jungle. The song, or chant, or whatever it was, took my mind away from the pain; somehow raising me above it. It was only when he paused to blow smoke or spittle in my face that it surged and gouged at me anew.

The icaro rose in volume and intensity and my gut cramped like a vice. Red-hot hooks pulled at my innards. I squirmed. Unable to ignore the agony anymore, I plunged down into it. I wanted to cry out but my voice was gone. Only breath remained. Then Don Sharaco fell silent and time stopped with him. Now even my breath was gone; my lungs were paralyzed and wouldn't move. I panicked. Blood pumped and pounded in my head but all else was still. The whispering rattle of leaves returned. Don Sharaco came in close, his face an inch from my stomach. "Haaaaarumph!" he said, and it was over.

I gulped air into my now fully-functioning lungs, so relieved I could have wept. The curandero bent and poked my stomach. The pain was gone. He grunted, patted my shoulder, picked up his things, and left the room without another word. The old woman smiled at me - her teeth impossibly even and dazzling white - then followed him out. Alf, who didn't even glance in my direction, went out after them, leaving me alone, stunned, confused and elated.

"You look better," Alf observed, when he reappeared a few minutes later.

"I feel better. Everything's gone. I can even breathe properly. What the hell did he do? He didn't even use any medicine."

Falling to Fly

Alf smiled. "You really want to know?"

"Of course." I raised a finger to my eye. It was still crusted and sore. Not a full miracle worker then. But close enough.

"Your eye is a different problem," Alf said. "But your bones and lungs feel better now, right?"

"They do," I agreed.

"There was a spirit inside of you, Jason. I sensed it when we first met. The spirit is very old and complicated, according to Don Sharaco. He used the words *duende* and *fantasma*, which mean goblin and ghost in English. But I couldn't understand his explanation fully. Don't worry though, it isn't malicious and it's happy to be not so entangled in you now."

"Right, okay. Good for the goblin. I'm just happy it's no longer twisting up my insides." I didn't know what else to say. I felt physically better, but also shaken to my core. "Thanks, Alf. I'm going to go rest in my tambo now. I'm sure this space needs to be prepared for tonight anyway, and I think I'd like to be by myself for a while."

"Of course. I will come when it is time for the ceremony."

I shook my head. "No. Thanks. I'm going to sit this one out. I've had enough mystical healing for one day."

Alf nodded. "You can't put it off forever though. You know that, right?"

"Right," I said, struggling to my feet. My legs were shaky but I managed to stagger back to my tambo without further mishap.

I slept through until evening. When I awoke it was dark outside, and eerie quiet. It was as if the trees and their inhabitants were holding their collective breath. I'd never known the jungle so silent, especially not at night. Then I heard Don Sharaco's voice, drifting through the air from the maloka. His icaro broke the reverential silence and within a few seconds the jungle came alive again.

I lay still, listening to the insects, the rustling of leaves and branches and the sounds of the ceremony. I had a sudden urge to run up and join them, but then I heard screaming and the unmistakable heave and splatter of violent vomiting and thought better of it. The moon rose, bright yellow and nearly full. A new voice floated down from the maloka, a woman's voice. Strong and pure as clear water. Both the icaro and moonlight lit up the jungle.

The foliage outside my tambo shook as an animal passed by. I sat up, straining to see what it could have been. Something big, that was for sure. I couldn't see it, but I heard it circling back. It was growling and padding back and forth. The wooden step leading to my door creaked and a deep barking roar turned my blood to ice.

The beast withdrew from the step and went back to stalking around the tambo. So thin and insubstantial were the walls, I could hear every rumbling breath, every snort and grunt. Minutes stretched into an hour. Then two. The constant tread and snarls, the swishing of leaves and tail, were hypnotic. After a while, petrified as I was, I couldn't help but fall asleep.

The trickle of adrenaline must have flowed into my dreams for within them I wandered outside and dashed through the moonlit jungle, sometimes running from, and sometimes with, a colossal jaguar. As I ran, an old woman sang in a cracked, bitter voice. I couldn't understand what she was singing but suspected it to be some kind of long and involved magic spell.

Every other living creature understood the power in the woman's song and they scorned me for my hateful ignorance of her message. Monkeys sniggered and chattered and screeched, birds squawked and beat their wings against my dumb head. Branches reached out from passing trees and slapped at my face, my arms, my back. I was adrift and stumbling when the sun rose and I was back on a familiar trail. The ground shook and the vines exploded through the earth for what I knew would be the very last time.

*

"Jason, come!"

I was still in bed. Everything outside was pale and glistening. It was early morning; earlier than it had been in my dream.

"Jason, come!"

I half recognized the voice but couldn't quite recall from where. Stopping to pull on a t-shirt and shorts, I opened the tambo door and saw Don Sharaco standing on the path. Ben and Alf were a few yards behind him.

The shaman pointed at the ground. "Jaguar," he said, raising his eyes to meet mine. "You hear it?" I nodded and the old man smiled. "You hear it and I feel it. All the night I feel it. Come!" He waved me forward and I went down the steps to meet him.

We followed the enormous pawprints around the tambo. There

Falling to Fly

were so many, but Don Sharaco was able to track their order. He showed us an area of flattened grass and mud where the jaguar had lain down at some point in the night. He pulled a few strands of black hair off a prickly stem and handed them to me. Tapping the wall of the tambo beside his head, he said, "Jaguar sleep by you. You know this?"

"No," I admitted. "I just heard it walking around out here."

Don Sharaco nodded then spoke in his own language to Alf.

"He said jaguars don't usually come here," Alf translated. "Last night was very special. He invites you to drink with him in the next ceremony. He says somebody who you need to meet is going to be there."

"At the ceremony?"

Alf frowned. "Yes, I think that's what he means."

Don Sharaco patted my shoulder like he had after performing his crazy exorcism the day before, then he walked away. Alf nodded to Ben and me and followed his mentor up the path.

Ben stared at the ground, wide-eyed and grinning. "A jaguar was sleeping outside our room while I was flying through space and time. What a night!"

I smiled too. "So how was it? Worth the money?"

Ben blew through his lips, making a sound like a horse. "You know, I'm not sure I could tell you what money even means anymore. It was horrible, and eye opening, and great. I puked, of course, and shat. But the whole thing was incredible. Not the puking and shitting - although that wasn't so bad, either - but everything else was, well, amazing really. Well worth whatever it is that's worth something, if you know what I mean."

"I haven't got a clue. But I'm guessing I will in a couple of days."

"You think you're ready to drink now?"

I rolled the jaguar's coarse hairs between my fingers and thumb and traced the outline of its pawprint with my toe. "I don't know if I'll ever be ready. But I think it's time."

*

The jaguar's nocturnal visit put me on high that lasted until about two hours before my first ayahuasca ceremony was due to start.

I had finished my day's work and was putting away my seed-cube maker when Tori turned up to tell me Yusuf wanted to see me. I was nervous about the ceremony and had hoped to spend some time alone to prepare but speaking with Yusuf took priority. I'd been trying for days to have him look at the plans I'd made for the gardens, and I also wanted to show him how well my makeshift nursery was coming along. The moringa was ready to plant, and so were many of the vegetable seedlings.

I found Yusuf sitting at the bamboo desk in his tambo. When he saw me, he closed his laptop and asked me to take a seat. There was a child-size stool placed across the desk from his beige leather swivel chair. I sat on it and peered over the edge of the desk up at him. My knees jutted to shoulder height. I felt ridiculous.

"Sorry about the stool, brother," said Yusuf, with a dismissive wave of the hand and a sympathetic smile. "Nothing else to use. Anyway, this shouldn't take long. So, how do you think things are going?"

"Pretty good. I've got the nursery up and running and we're ready to start planting." I placed the sketches I'd made detailing the garden plans on the desk. "Once you check these out, of course. I was thinking to start on the land behind the kitchens. It's not being used for anything else and permaculture principles - along with common sense - suggest having the kitchen-garden right by the kitchen. What do you think?"

Yusuf pushed the grubby papers aside. "I'll check them out later. Look Jason, I'll come straight to the point. I'm disappointed in you and I don't think our arrangement is going to work out. You've been sick ever since you got here. You've set up a makeshift nursery on land we need for a tambo, without asking anyone if it was okay. And your energy is just wrong. I should know; reading energies is something I'm very good at. I almost became a shaman, you know."

"Wow. Really?" If he caught the sarcasm in my voice, nothing in his bland expression showed it. "I know I've been sick, but I've still been working. And as for the nursery, it can easily be moved when you're ready to start building. I *did* try asking you about it, but you've had no time to see me until now."

"You're saying it's my fault?"

"No, not entirely. But we should be communicating more if we're going to make things work."

"You're not hearing me, Jason. We're not going to make things work. Bottom line is you're too needy and inconsiderate. I know you took the whole afternoon off the other day, again without asking permission."

"I was sick. Ask Alf and Don Sharaco. He had to do a healing on me."

"Yes, I heard about you wasting our shaman's time."

"Did he say that?"

"No, but I know that's what happened. I also heard you had some kind of trauma back in Ecuador, which has clearly messed you up. We don't need a gardener with post-traumatic stress disorder. You'll destroy the nutrients in our food with all that bad energy."

"Are you for real? I told you what I'd been through when I first wrote to you. I thought that was why you agreed to our trade."

"No, I agreed because I needed someone to set up the permaculture side of things. I don't have time to read sob stories."

"It wasn't a sob story, and I'm not suffering from post-traumatic stress. I just wanted to create your gardens and maybe do a ceremony or two. I'm sorry if you don't think it's working out, but so far I've kept my side of the bargain." I nodded at the pile of drawings on his desk. "Those are detailed year-by-year plans for the next fifteen years. You can see the state my eyes are in. It wasn't easy drawing them up."

Yusuf glanced at the top sheet. "Clearly. They're scruffy as hell. Illegible." He leaned back and swiveled side-to-side in his chair. "Okay, Jason. In exchange for the worms and other stuff you brought here and the work you've done so far, and, say, two more weeks to train and explain these plans to our workers, I'll let you do the ceremony tonight and the sanango cleanse with Alf afterward. But to be honest, I don't think you'll be able to handle it. I don't think you're strong enough."

"Not strong enough for what? The ayahuasca or the sanango?"

"Both. Aya can be harsh on people like you. And the sanango takes five days of deep commitment. You'll think you're dying and you'll quit. That's how it'll be."

These were the first details I'd heard about the sanango cleanse. "What happens during the five days?"

"I don't know exactly. I've not done it myself yet. But I do know it'll be too intense for you, brother."

"Brother? We're not brothers, Yusuf. You don't know anything about me. You're too wrapped up in yourself." I gripped the edge of his desk and pulled myself up until I was standing. "I'll be strong enough for the ceremony tonight, and I'll be strong enough to do what you're too afraid to do. In return, I'll stay another two weeks and train your workers how to follow the plans I've made. You'll have great gardens, and every time you eat from them you'll remember how wrong you were about me."

"Deal," said Yusuf. "You can go now."

I walked back to my tambo feeling furious, not to mention sick and lost and scared. In spite of my bravado, I was terrified of drinking ayahuasca, and now I was worried about the sanango too.

The closer I got to the ceremony, the more stressed I became. I was worried the ayahuasca would turn me into a deluded fool, like Yusuf and Tori, and I was also worried about what would come next. I'd committed to being in Peru for months, but now I'd have to go home with my tail between my legs. All in all, I couldn't have been in a worse frame of mind when Ben came to tell me it was time to go up to the maloka.

*

When we arrived at the maloka, there were wooden half-chair recliners without legs, yoga mats and cushions, all arranged in a circle around the edge of the ceremonial space. In the middle of the room a brazier smoldered, filling the air with fragrant smoke. Next to each chair was a plastic puke bucket.

Ben and I took our places next to Don Sharaco, who seemed to be in a state of deep meditation and hadn't noticed us come in. Across the room I saw the old lady who'd been with him when he'd performed the healing on me. She smiled and waved. I waved back, and then sat in silence while the room filled with our fellow participants.

When Alf came in he sat on the other side of Don Sharaco. Tapping the old shaman's arm, he leaned in and whispered something to him. Don Sharaco nodded, then turned to look at me. He smiled, and I instantly felt a little better.

This almost but not quite hopeful feeling continued right up until Yusuf and Tori came in. Yusuf didn't look over but Tori did. Her top lip curled into an Elvis-like snarl at the sight of me. Somehow, I'd thought, or hoped, they wouldn't be there. Their presence made me more nervous than ever. I stared at my plastic bucket, wondering if I'd spend the entire evening with my head wedged inside it.

"Don't worry," whispered Ben. "I know you weren't following the diet until you got here, but I don't think it matters. I spoke to Don Sharaco. He said it's good to cleanse, but normally he has fried chicken and yucca before a ceremony and he hasn't had to purge in over twenty years! So much for this gringo aya-diet bullshit. You may not need to puke at all."

I couldn't help but smile at him. "Thanks Ben. I'm glad you're here."

He grinned back. "Glad you're here too, man."

Don Sharaco stood up and cleared his throat and all the hushed conversations around the room fell silent. He stepped forward toward the brazier and began an icaro. Flames flared and then died off. Smoke clouded and swirled. Don Sharaco shook his bunch of leaves in a disjointed rhythm and, one by one, we went up to the brazier. Each of us followed Alf's example, wafting smoke over ourselves while the shaman moved around behind us, shaking his leaves and singing.

Once this cleansing ritual had been completed it was time to drink. Because I was sitting to the shaman's left, I went first. I had no idea how awful ayahuasca would taste until then. Swallowing the vile-tasting brew is an ordeal unto itself. It's sickly sweet and sour and bitter as bile. Don Sharaco indicated that I down it in one. I did, trying not to grimace and trembling all the while. The aftertaste was like a mix of old tobacco and coffee grounds mixed in syrup.

I sat back down and watched everyone else drink. It was funny seeing their expressions. No matter how noble or crude their reasons for drinking, the taste got them all the same way. All except Yusuf, who made a big noisy show of how bad it was. I noticed Don Sharaco look away from him, so as not to witness the disrespect. Or at least that's how I saw it.

We were asked to set a personal intention for the ceremony. If there were any questions we wanted answered, or any healing or advice we wanted to receive, now was the time to focus upon it. I closed my eyes and tried to think of something.

Show me what I need to see, but please be gentle on me, was what I came up with.

Don Sharaco began another icaro. This time I felt the flow of it in my gut. I still had the aftertaste of the ayahuasca in my mouth and in the back of my throat, where it thickened to sludge. I fought back the urge to gag.

"It's time," said a woman's voice.

I looked about but couldn't see who'd spoken. "Time for what?" I asked aloud.

"Let go," she said from behind me.

I whipped my head around and strained to see where she was, where the voice was coming from. The more I searched, the more squirmy and nauseous I felt. "What do you mean?" I asked. Bubbles of bile popped at the back of my throat in answer. "Oh. You mean time to purge, right?"

I didn't wait for an answer. I grabbed my sick bucket and threw up. A charged liquid fizzed through my body and whenever it hit something it didn't like, it took it out - pushing it up and out through my stomach and throat and mouth. Lumps of black, sparking like coals, spattered into my bucket. I vomited seven times in total. Each time, I felt a little better; cleaner, readier. By the time I finished, my body was buzzing and felt as if it was running at a hundred-percent physical efficiency.

I leaned back on the half-chair and closed my eyes.

Serpents advance toward me from the darkness. They are huge and they intertwine between one another at a bewildering speed. Slithering, shuddering, scuffing. They buzz and crackle like electrical power lines. As I think this, the snakes glow neon and hum, clicking and twirling, their secret unlocking, to form a shifting tunnel of unfolding geometric perfection.

The tunnel pulls me in and I fly, my arms out in front like superman. Sometime fast, sometimes slow, I glide onward, cutting into sharp bends and sweeping round gentle curves. I have lost all sense of time or place.

Without warning, I'm back in Texas, in my mother's living room. She hands me coffee. Our eyes meet and we switch perspectives. Now I see me. Pathetic. Pitiful. Sitting on her couch with my excuses. Thirty seconds of this and then time jumps; the frame shifts. It's a week later. She watches me leave. Why can't you settle? What did I do wrong with you? Why can't you be normal?

I'm at my sister's house now, and I'm back to being me. She's younger. No kids. When are we? Now I'm her, seeing me. I don't understand you, Jason. You're my brother but I just don't get what makes you tick. Why can't you find someone and settle down like everybody else? Why are you so restless? You're such an embarrassment!

Further back. Oh God, no. I'm in Presa River. Is Gwen still here? Tears blur my vision. My heart aches with missing her, even after so many years. I see her now; so beautiful and young, so full of promise and love and dark secrets, all mixed up and churning inside. She sees me. You're not enough to keep me here. *Am I thinking that, or is she? A week later and she's gone. Must have been her. No time to cry, but I cry anyway. My ribs constrict and my heart sinks into my belly.*

Alf is here. No, not Alf - Alf's eyes in a frail boy with curly dark hair. He sees more than me. He sees the shadow. My breath catches. I see it too. It's so much bigger than I am.

The clips speed up. I'm spending less and less time in each place. My dad flies by in seconds. I want to stay with him, but all I feel is his disappointment and then he's gone. I'm in Alaska. Oregon. Belize. A moment, a week, a month, a year later, or before. I'm with old friends, family, girlfriends. All are disappointed in me. I'm too sad, too weak, too distant, too weird, too much. My guts whorl and I heave and wheeze.

The images spin faster, and grow darker, more sinister. I'm in the ruin. *Mikey admires me. I'm brave. No, I'm lost.* What's wrong with you? What are you trying to prove? Doesn't matter how many times you go in, you're still a loser! *The face is there in* the ruin. *An old man with curtains of white hair and an evil stare. The sweet, cloying smell of death. I can't escape it. I never could. I retch and belch the foulness of rotting flesh.*

His hatred burns in the eyes of another face now. This new, younger man hits me over and over. My eye socket explodes in pain, yet he feels nothing but a detached animosity. I know it because I am him until I flip back to me. Someone else is strangling me. I can't breathe. He doesn't care. He wants me dead and gone. My heart is breaking with pity - but pity for whom? I'm the one in pain, but I stopped caring about me long ago. I gasp for air and open my eyes.

*

Back in the maloka, Don Sharaco senses my distress. He's been there throughout. He changes the icaro and sends waves of light toward me, pale blue and sparking. His singing continues but his face stops and he turns toward me. He nods. I nod back. I'm doing okay.

*

The most beautiful sound fills the maloka now. I don't know what it is. I'm not sure if it's a human voice or some exotic stringed instrument. Then I see her: Doña Luisa - the old woman with the too perfect teeth. She sings the icaro and it's so sweet my chest swells and my heart rises and falls in joyful syncopation to the rhythm. I want to weep tears of delight, but all I can manage is a chuckle. I look again for her face across the room. There she is. Still as a statue. No, not still - she's rippling; her mouth and throat and stomach undulate, calling the icaro up from deep within the earth.

Beside her sits an old man. I hadn't noticed him before. His hair is long and parted in the middle. Blonde or white, I can't tell. He looks up and sees me watching him and a jolt of terror shudders through my bones. It's him: the demon from my childhood. We switch perspectives and I see myself - a small, pathetic child, seated among his people; white demons who live for the disease that consumes and corrupts.

I reel backward into darkness again. Pixelated lilac and lime serpents come at me, mouths agape, fangs dripping venom. I open my eyes, searching for Don Sharaco.

He's not here. No, it's me who's not here. I'm walking in the sun of another time, in a permaculture-paradise beside a river. The world is younger and I am someone else. I'm at peace, and I'm warm from the sun upon my skin. There's no disappointment anymore. I feel only admiration, love, respect.

Canoes glide by below. Otter Tribe. A blue-black serpent rises out of the long grass. Our eyes lock and I take off, slithering into the water.

From the river I see the pale demon-men with thick curly beards being worshiped as gods. Arrows come in fast from the shore. The Otter Tribe falls and splash into spreading clouds of red.

The demon-disease of ownership and insatiable greed is coming for us and I'm afraid. The head-demon has a huge eagle with a crested head in a cage. Within the eagle's fire-wreathed eye I see more suffering; far off people enslaved and flogged. Everything hurts. Why? I ask, and I see the familiar gleam of gold, a vast building - a church or castle made of blood and mud and stone and colored glass.

I'm confused and sick with anger and grief. I don't want it here and I run. But it washes in like a flood, covering my land in discontent and fear. Nobody listens now everyone is infected. All are guilty. Even me. Because I see it and talk of it I am hated. Tears are streaming down my face. My heart is shattered, my mind slips into madness.

What's happening to me? Where have I gone?

I hide away underground. Knives flash in firelight. There's pain in my belly. My back. My side. Water splashes then carries me into darkness. I drift and die. My spirit released, stays in the flow of the dark river. I whirl and churn in a world of shadow and feeble misery.

*

I'm shivering and crying and back in the maloka but I don't know who I am.

Am I Jason or Casqui the mad man? Or both? or neither? If I'm not Casqui, how do I know that's my name? *I'm so confused. My head is pounding. I don't know what's real or if I'll ever be sane again.*

I grip the plastic bucket. Doña Luisa is still singing but the man is no longer beside her. I'm sweating and so cold and I vomit again. The sparkling lumps I puked earlier are gone; the bucket is almost empty. Doña Luisa's icaro shifts in tone and her hand reaches out to me. She's on the opposite side of the room, but I reach back to her. Our fingertips touch and I am transported to a quiet place of honeyed light.

Questions! I need to ask questions! *I don't know how I know this, but I do. I see flashes of my earlier visions and wonder if I should be more responsible; go back to the States, settle down; make my parents happy.*

Have I been a selfish person in my life? *I ask.*

The voice that answers is my own. Yes, *it says.*

Though it sounded like me, I'm sure it's really the spirit of ayahuasca. Which is good, as I have something else on my mind, something that's been bothering me for a while.

Was it my fault I was attacked? Did I bring it on myself?

No. *The answer is emphatic and a great relief floods though my body.*

Spurred on by this release, I ask, Is life about unconditional love, family or something else?

The answer comes in the form of an intense embrace. I see my mother and father smiling at me. Alongside them are Tyaento - the Huaorani I hate - and the old man - for he's only a man now, tragic and embittered - from my haunted childhood, and a beautiful little girl with cornflower-blue eyes who I don't recognize although she's as familiar as my own reflection, and a crowd of other people, and animals, and birds. Everything is turning and thriving, flowering and flowing. I'm overwhelmed by it all.

I hear a woman say, You should smile more.

I smile. I can't help it. The smile turns into a big yawn and my ears pop and I hear the sounds of the jungle beyond the maloka.

You should smile more!

This time the woman's voice booms and reverberates inside my head. I smile and yawn again. And once again, my ears pop and clear. This time I hear every living thing in the vicinity. Breath and scratch and breeze, everything to the smallest movement and vibration.

You should smile more!

This time I try not to, but it's impossible. I have to smile and I do. The same sequence happens over and over until a profound serenity settles over me. From it comes a sudden giddiness and a compulsion to tell everyone how wonderful I feel.

I love you, *I say to Don Sharaco.* I love you, *to Ben and Hedgy.* I love you, *I call out to Doña Luisa, and to Yusuf and Tori. I keep going, making sure everyone in the maloka knows they're loved by me. Nobody hears me but it doesn't matter. They're all laughing and smiling, yawning or mouthing silent proclamations of their own. We are sharing this moment in a paradox of solitude and connection. It is as it should be.*

*

Content, I lie flat and close my eyes and I'm flying again. This time gliding over a jungle river, following its meandering course. Vines are racing up the trees and cliffs fast, as if captured by stop-motion cameras. I try to move my head, but I can't; I'm not in control here; I'm an observer, nothing more.

I gain altitude and soar over the canopy. In the distance, I see a structure. It's high-tech and metallic, rounded, immense. Only small portions poke through the treetops; tips of the vast iceberg; a civilization frozen in time. No, not frozen. Gone forever. There are no people left. Soon the metal rusts and disintegrates, leaving only hauled and hewn stone blocks behind. The jungle continues to thrive as if nothing happened. A monkey laughs from atop a pile of stones.

*

I opened my eyes and saw people milling around. The light was brighter than before. Ben was sitting in silence beside me, eyes closed, hugging his knees into his chest. Hedgy smiled across at me.

"You two are the last to come around," she said, nodding at Ben. "The ceremony finished a while ago. How are you feeling? Did you have a good one?"

I opened my mouth, then closed it. There was nothing to say.

Sanango

Alf came for me early in the morning. I hadn't really slept after the ayahuasca ceremony. My mind had wandered, replaying the images and visions over and over, wondering what to make of it all. I felt raw and turned inside out. Exposed and vulnerable, on the verge of tears, yet manic, exhilarated and exhausted.

Ben had slept from the moment we'd returned from the maloka and he slumbered through Alf's gentle knocking. I pulled on some shorts and a t-shirt and went outside to see what was up. Once we'd found a place to sit along the pathway, Alf turned to face me.

"How are you doing?" he asked. "Did you have a good ceremony last night?"

I shrugged, still not knowing if I was able to speak yet.

He nodded. "Tonight, you'll be meeting the grandfather, and I want to be sure you are ready. That's why I came so early. It will be different to ayahuasca. The journey is not really psychedelic, but it's still very challenging. We should talk about it now so you can decide and prepare."

I knew he was talking about the sanango cleanse, but for some reason I played dumb. "Grandfather?"

"Yes. Uchu sanango is called the grandfather. You'll find the name is appropriate to the medicine. Same way ayahuasca is called the mother."

Although I'd heard ayahuasca referred to as the mother, I hadn't felt anything very maternal in the experience, so I wasn't sure what to expect from sanango based on its familial nomenclature. I considered the only grandfather I'd ever known; the sour and spiteful old man who'd raised my mother and whose apartment we'd lived in when I was a small boy. The wounds of his resentment and disappointment had been torn open afresh during the visceral visions of the previous night, while clutched in mama ayahuasca's nurturing embrace.

"Not all grandfathers are the same," Alf said, once again reading my thoughts through my facial expressions. "This one is strict but also very wise and has your best interests at heart."

"Okay, so what do I need to do to prepare for it?"

Alf smiled. "At the moment, nothing but rest. We'll begin tonight. After dinner I'll take you to the sanango tambo where you'll be spending the next five days in isolation. I will be your only contact with the outside world during the cleanse. I'll bring the medicine and perform the ceremonies. You'll be drinking the sanango every night for three nights, and then you'll have two days to return to yourself. I'll also be bringing you food and checking in on you, although you probably won't notice me doing it."

I raised my eyebrows at that. "Why?"

"Because you have to eat. Speaking of which, the dieta is very strict and you have to commit to it for the next six months."

"I meant why won't I notice you coming in? Hold on, six months?"

"Because you won't be interested in me or what I'm doing. You'll have other things on your mind. So, yes, that's right - no pork or red meat for six months. No spices, salt or oil for a month. And no sex or masturbation, or even thinking about sex if you can help it, for the first month as well. Also, no sharing of bodily fluids with anyone - meaning drinking from the same water bottle, that kind of thing. Oh, and no products like soap or shampoo, either. Don't worry, I'll write everything out. You won't have to remember it all."

"Good job I don't have a girlfriend or she'd be pissed," I remarked. "I can't even think about sex, huh? Is that one serious?"

"It's all serious, Jason. These medicines are not to be taken lightly. You could really hurt yourself if you don't adhere to the sanango dieta, during and after. If you can't commit wholeheartedly, it's better you don't do it."

"But you said I needed to do it."

Alf looked away. "Plenty of people need to do things but don't end up doing them. They're too afraid or weak or preoccupied with what others might think or say."

I touched his shoulder and he turned back toward me. "I'm not one of those people," I insisted. "I'll see it through."

His pale green gaze pierced all the way to the back of my skull. "I know you will. You're a lot stronger than you think, Jason. Besides, when you feel weak, the grandfather will help you through. See you this evening."

As he walked away, a rush of questions tumbled through my mind. I tried to hold them, to keep them for later, but they slipped away like grains of sand in a clenched fist. I shook my head and decided to go for a walk in the forest.

I didn't walk far. The effects of the ayahuasca were still in my system, making everything I saw a wonder. I stopped at every shard of light, every veined leaf, each dance of butterflies, all the luxuriant carpets of moss, luminous and hugging the damp, dark wood of their home. My love of nature made more sense than ever now. The connection was real and so holy it put all religions and their accoutrements, their dead relics, to shame.

When I returned to the center it was lunch time. Everyone appeared softened and content; basking in the afterglow of the ceremony.

"Hey Jason," said Hedgy, handing me a plate of unsalted rice and steamed river fish. "Alf told me to give you this. You're starting the sanango cleanse tonight, right? Hardcore, man!"

I smiled and took the plate. "Thanks, Hedgy. How was your ceremony?"

"Difficult. Much more difficult than the first. I had some scary shit go down, but I guess it must have been necessary. Today I feel amazing. Really open and light and grateful. How about you?" She grinned. "You didn't seem too talkative afterward."

"It was good, I think. Lots of questions were answered but just as many new ones came up. I'm still pretty shell-shocked by it all, to be honest. I'm not feeling much of an afterglow. I just feel all scraped out and jittery."

"Man, you sure you're ready for the sanango? I heard it's tough."

"I might be if people stopped telling me how hard it's going to be."

Hedgy chuckled. "To be forewarned is to be forearmed."

"What?"

"It's an old-fashioned way of saying brace yourself because shit's about to go down whether you like it or not."

*

After lunch I tried to nap in one of the hammocks scattered around the grounds. I chose the most secluded I knew of and laid back and closed my eyes. The back of my eyelids lit up in vivid shifting colors, like a kaleidoscope, with circular patterns forming, falling away and reforming again. I couldn't sleep but found the odd private light show of the residual ayahuasca calming enough to feel pretty well-rested afterward.

At dinner I said goodbye to Hedgy and Ben and all the other participants I'd gotten to know during the retreat. They were staying for one more ayahuasca ceremony but would be gone by the time I was due to emerge from the sanango cleanse. Tori and Yusuf took their lunch and dinner in their tambo so I didn't get to see them all day, which was fine by me; I felt fragile and nervous enough without having to deal with them as well.

When I was ready, or as ready as I'd ever be, Alf led me away from the maloka and down a long path to the tambo that would be my home for the next five days. It was the most remote of all the tambos and was used only for sanango retreats. A stream trickled by outside, feeding a deep pool of clear water to one side of the door steps. A bucket containing flower petals was placed beside it.

"Your bathing hole," said Alf, nodding to the pool. "The flowers are medicinal and work with the sanango during the cleanse. Pour them over your head whenever you take a bath, which you'll probably want to do at least four or five times a day. As you can see, the water is clean and pure and constantly flowing in and out. You can settle in now. I'll be back at three and we'll begin."

I turned on the light and inspected my new room. There was a bed with laundered sheets and a light blanket, a wooden table and chair, and a hammock. There was a toilet and wash basin set back into a small alcove. A shelf above was stacked with toilet rolls and three neatly folded threadbare towels. No reading materials or other diversions; this was a Spartan place of healing and nothing more. I stretched out on the bed and listened to the sounds of the jungle and waited for Alf to return.

It was full dark when Alf arrived at three in the morning, carrying a tray with everything we would need for the ceremony. He lit a loose-rolled cigar of mapacho - sacred tobacco - and puffed it like a train, exhaling all the time into the cup of sanango I was meant to drink. His level of concentration and intent were mesmerizing. Out of nowhere he raised a bunch of leaves, shook them, and began to sing an icaro.

As he sang he wafted the mapacho smoke over me with the leaves, much the same as had been done at the beginning of the ayahuasca ceremony and during the weird healing Don Sharaco had performed in what now seemed a lifetime ago.

Alf nodded to the cup and I raised it to my mouth to drink. It tasted like smoke and burned like fire all the way down to my belly. The warmth radiated slowly outward as Alf continued to chant and puff and shake his bundle of leaves. My body trembled in time to the leaves until we were so in synch they appeared to stay still while the whole world juddered around us. The waves of heat grew into an intense fever. I dropped to the floor and curled up, shuddering and sweltering in a pool of my own perspiration.

My vision was blurred coming in to the ceremony as my eyes were still too infected to wear my contacts. Now it was even fuzzier, dimmer, more indistinct. I blinked, expecting it to clear, but instead I was plunged into utter darkness. Blind and helpless, I called out to Alf.

He was already gone. I'd heard him pack up and leave, though, as he'd predicted, I'd thought little of it at the time. But now I felt abandoned and alone and scared to death something was going wrong. I tried to call in the spirit of sanango, the wise and benevolent grandfather Alf had claimed would take care of me. In return I heard a deep throaty chuckle and malicious whispering voices.

"Who's there?" I demanded.

Silence. I blinked and was blinded by a bright cold light. Freezing now, I shivered and my teeth chattered. The light was everywhere. Soft and cottony like a cloud. I was in another dimension, and the thought of it gave me an odd sense of comfort, as if the normal rules of being no longer applied and I was safe from the chill torturing my body.

Later, at the very moment I felt I might actually die of exposure and be frozen to death in the jungle, darkness fell again and with it came the searing heat of the fever. This cycle continued for hours. Dark and hot. Light and cold. Then dark and cold, light and hot. It was discombobulating and uncomfortable, but I was past caring about it.

Eventually the light won out over the dark and I assumed morning had arrived. There was a smear of thick film over my eyes and no matter how much I tried to wipe it away, it wouldn't budge. I was still racked by chills and fevers, but my vision no longer lurched between dark and glaring light; I could make out shapes now. The putrid smell of my own body disgusted me. I needed to wash. I remembered the bathing hole and crawled toward the tall rectangle of light that was the open doorway of the tambo.

I slithered down the steps on my belly and rolled into the water. It was warm, womblike, energizing. I fumbled around the edge until I found the bucket of medicinal flowers and upended it over my head. I gasped at the icy shock of the cool water and the fragrant pungency of the flowers. My mind swam and my body bobbed back and forth. I blinked and was once again in darkness. I held my breath and lifted my legs, hugging my knees to my chest and allowing my body to tip and sink beneath the surface. All I could hear was my heart beat and the swish of water passing my ears.

When I looked up I saw a circle of light and felt as if I was deep inside a well, secure in a cozy world of limited vision. I stood and my vision was flooded with verdant pastel shades, diffused and expansive. I hauled myself out of the bathing hole and made my way to the bed where I collapsed and fell asleep, my dripping body wrapped in the blanket.

I drifted in and out of consciousness throughout the daylight hours. My mind wandered and my body trembled, sometimes in fever, sometimes with chills. Whenever I could, I ate, or went to the bathing hole to wash.

Time passed normally with sanango and I had no real visions - the journey was physical and mental and soul deep; shifting my reality in ways ayahuasca hadn't reached. I felt as though my perspective was being broadened and sharpened and turned inward, then outward, in waves of harsh clarity. I was aware of the shortcomings and depravity I held within and how my thoughts and actions rippled into the wider world. Sanango seemed to be urging me to be a better man, whatever that might mean.

When darkness fell, I stopped thinking and gave myself up to the harsh noises and inky blackness of the night. Alf appeared at some point to smoke mapacho, sing icaros and give me my second cup of smoldering sanango. I wondered if this was how dragon's saliva, or more likely dragon's puke, tasted. My mind drifted to thoughts of dragon's blood and its magical healing properties, and I hoped it and the sanango were somehow related. I began to shiver and spasm anew. This time I made it to my bed before I was blinded.

I was becoming accustomed to the cycle of fevers and chills, but it didn't make them any easier. Once again, I heard voices mumbling somewhere nearby. I strained to understand what was being said, but the moment I grasped a word or phrase, it slipped away. It was like trying to make sense of someone talking half-gibberish in their sleep. In the end I gave up and let the hum of the phantom conversations fall away and blend into the white noise of the jungle night.

I had the preternatural awareness of traveling through other dimensions again, while not ever having to relinquish my foothold in my own. This gave me a temporary ability to contemplate things from angles never before considered. Over and over, throughout it all, one word kept flashing in my mind: Integrity.

I wasn't sure why I kept returning to this word and its meaning. It struck me as a powerful virtue often pushed aside in favor of inferior motivations by almost everyone I knew, including myself.

This would have to change if anything was to improve in the world, I decided in my rarefied state of elevated consciousness. Because of what happened next, I suspected this focus on integrity may have been something of a parting gift; the body of a lifetime's contemplation distilled and bequeathed in the hope it might make it through my thick skull and give me some well-needed direction.

A surge of energy pulsed through me, and I jumped up and yelled my own name in someone else's voice. The sound was guttural and booming, with a vitality all its own. It hadn't come from me, that was for sure. I was shocked by the strangeness but felt unafraid. Maybe it was the voice of sanango, or perhaps some other swirling spirit of the jungle. The fever ceased immediately and a pleasant vibrating numbness spread throughout my body.

I laid back down. All my senses were heightened and I felt supercharged in the same way I had while climbing the cliff after my attack. The only difference being, back then I'd had to keep moving in order to survive, whereas now I recognized my survival depended upon surrender. So, I let go.

My physical body fell away and I could no longer move. I lost all connection to my muscles, nerves, bones and breath. I was a soul floating, all sensation, pain, irritation gone. There was no sound, no vision, no light, no dark. No itch so nothing to scratch. It was unlimited freedom and I loved it there.

At least thirty minutes passed with me marveling at this new state of being; so innocent, so joyful. *This* is how we arrive in the world, I realized. Before we're molded by our societies, this is what the raw material feels like. If only we remembered and let it inform our behavior and cultures instead of allowing it to be snuffed out at the earliest opportunity.

The feeling of vastness and disembodiment ended abruptly and I zipped back into my meagre meat suit. I was disappointed, yet not entirely unhappy. Something had changed while I was gone. I felt lighter. Whatever had been festering inside me for so many years had been released.

I thought of *the ruin* to see if I could conjure the feeling of the watchful entity, knowing already that I couldn't. I saw the face of the old man in my mind's eye, but the hatred and bitterness were transformed into a dignified pity, tempered with a tiny gleam of hope. Whether the entity had been real, as Alf had insisted, or merely a manifestation of my damaged psyche, didn't matter. It had now departed. Gladdened and exhausted, I fell into a deep and restful sleep.

When I awoke, I crawled to the bathing hole and scrubbed myself with the flower petals before dunking my body into the pool. I was clean and refreshed and felt I'd achieved what I'd come for. I was ready to return to the world I'd left behind. It was time I settled down and paid more attention to my loved ones.

Not only did I want to be there for them now, but for the first time ever I felt able. My heart was full of love and vulnerability and forgiveness, and my will was tempered with new courage and resolution. I was committed to raising my integrity so I could become a better human being.

Alf came in the nighttime and I drank my final cup of sanango. The blindness intensified and I fell into the sequence of hot and cold, fever and chills, once again. This time I was grateful for it all. I knew I was being scoured out, cleansed and humbled in order to be fortified, rebuilt and rebooted.

It took thirty-six hours for my eyesight to return to normal. Alf came on the second-to-last night and sung icaros and blew mapacho smoke over me while I sat wrapped in my blanket, not seeing anything but black shadows dancing over my eyes. After the closing ceremony on the final night, Alf and I shared a smoke of mapacho in silence. When we finished he rose and told me he would see me at breakfast.

"Will there be anything other than rice and boiled eggs?" I asked, referring to the basic and monotonous fare Alf had supplied me with for the duration of the five-day cleanse.

He grinned. "Maybe some steamed vegetables for you, my friend. As a reward."

"Thanks." I smiled back. "I mean it, Alf. I appreciate everything you've done for me. And steamed veggies sound like manna from heaven right now. I don't think I could look at another boiled egg for at least a year."

"That's too bad. You need the protein and no one's gone fishing yet." He nodded to acknowledge my thanks. "You're welcome, my friend. Did you feel it leave?"

"I think so."

He scanned the area all around me. "It's free now. The last tether was severed by the sanango and you both now have a chance at peace. I hope you can find it."

*

I returned to the gardens after breakfast and spent the day peacefully tending the seedlings I'd germinated before the ceremonies. Yusuf and Tori weren't around. They'd flown to Bali to celebrate the success of the center's first retreat and wouldn't be back for six weeks.

Alf and two of the workers, Shara and Guillermo, joined me in the afternoon to begin their instruction in permaculture principles, and to help map out the sites for planting. I enjoyed their company and the focus teaching brought me, not to mention the bubbling stream of endorphins released into my system by the craft of gardening itself. I was calm and content and felt like a new man. The mysterious jungle medicines had worked their magic on me, far exceeding my expectations and hopes going in.

Of my three students, Shara was the most enthusiastic and attentive. He was in his thirties, about five feet tall and from a tribe that still lived, for the most part, deep in the jungle. Although he spoke perfect English and Spanish and seemed very westernized, his parents had never even been to Iquitos. They still practiced the same wild-crafting and basic agricultural skills that had seen their tribe thrive for millennia and had no interest in changing their ways. I was fascinated by his stories of hunting monkeys with blow pipes and setting pools fed by eddies with a potion made of crushed seeds to catch fish. One day, while we were preparing soil with the worm castings, I asked him why he'd left his tribe.

"I haven't left them," he insisted. "How is it possible to ever leave one's people?"

"Well, you're not there with them anymore," I observed.

He regarded me as if I were dense. "Yes, I am. And they are here with me," he said, touching a fist to his heart. "Is it not the same for your people? Do you not keep them with you?"

"I guess so. But what I meant is why are you physically so far away from them?"

He shrugged. "We travel for miles when hunting and we're often gone for many days. It's right for people to keep moving. It's why we have legs. We cannot always walk on the same paths either. If we did our prey would know when we were coming and we would not eat."

"But you're not out hunting. You're learning about permaculture with me and collecting a wage from Yusuf."

"It's still hunting," he chuckled. "Just not for monkeys. Here I'm hunting knowledge and money."

"To take back to your tribe?"

"Yes, of course." He paused, then added, "Not the money, though; I'll buy things with that. But the knowledge I will share, even if it makes them all fat!"

I laughed. "Why would it make them fat?"

"You told me when the gardens and food-forests are set up there will be a constant supply of good things to eat, all in one place."

"So?"

"Most jungle people eat what there is, when it is there. It works well out in nature because we have seasonal rhythms directing our lives. We have lean times when food is scarce and times when there is abundance and we can feast. If there is food all the time it is obvious we will get fat." He pulled his cheek back to reveal gold molars, top and bottom. "You see these?" he asked, tapping them. "The dentist told me my teeth rotted out because I drank nothing but Tang and Inca-cola and ate spoonful after spoonful of sugar when I first came to Iquitos. The thing is I love sweet things. In the jungle it never becomes a problem because honey is hard to come by. But in Iquitos there are shelves full of sodas and white sugar in every store." He shrugged. "If I'd stayed in the jungle I would still have my teeth like my father has."

As Shara spoke, something clicked in my mind. I'd always thought the Huaorani were disrespectful and greedy because they'd gorged themselves on our food until it was gone. It'd been one of the reasons my friends and I had resented them so much. But from what Shara was now telling me it could all have been a misunderstanding; a cultural difference in attitude, neither their fault nor ours.

"So, why did you leave the jungle?" I asked.

"Curiosity, at first. Also, for survival. We heard how the forests were getting smaller and how the rivers were being poisoned. I wanted to see how the people doing these things survived, so we could know how to do it, too. But I stayed in Iquitos because I became greedy. I saw more and wanted more than my life in the jungle had offered." He gave a dry laugh. "I refused to believe that having too much of what I craved was possible. But it made me sick in my body and mind and rotted my teeth. It will be the same for all my people in the end. We are not made to live in the western way."

"Some are," I said, thinking of Tyaento. "But they're the worse for it."

He nodded slowly. "So why did you leave your land, Jason?"

"Same as you but in reverse. I left because I wanted more. I was greedy for experience and gold. Now I've realized it's more important to be there for my people."

His eyes narrowed. "I hope that doesn't make us enemies one day."

"Why would it?"

"Because your people are killing mine and one day we will have to make a stand and fight back. If we give up and try to live your way, we will die. This is the most important thing I've learnt out here."

"And the permaculture?"

"Imitates nature, like you told us. It's good use of technology and knowledge for making things better. Don't look so sad, not everything your people think of is bad. You just dream the wrong dreams, and we all suffer for it."

"So, you plan to fight the guns and heavy machinery with blow pipes and poison?"

"If it comes to it, yes. Or maybe we'll think of some other way. Or you will. Perhaps you'll change your dreams and stop imagining you need oil and gold and the wood from our trees and leave us in peace."

"Fat chance of that, Shara. There's too much profit in it, unfortunately."

"Yes, and we are not immune to that desire either. We can be just as short-sighted as you. In some ways, more so, because we know so little of the outside world and the irreversible damage already done elsewhere. I hope my people will listen to me and heed my advice about the dangers of your ways, but it is far from certain they will. In fact, it's more likely they won't. The companies are very seductive in what they offer for land and labor. Still, I have to try. This is why I like this way of making gardens you're teaching; it will provide me with something solid to offer them when I return."

I smiled at that. "I hope it's enough, Shara. I really do."

Spending time with Shara opened my eyes, and also filled me with shame and despair. I'd never been one for white guilt. I always thought it dumb to assume responsibility for the crimes of my ancestors or others within my culture. But now I wasn't so sure of my lack of culpability.

After all, I drove a car when in the States, and used so many products that are derived from oil industry plastics. I also probably wiped my ass with paper made from rainforest trees without giving it a second thought. And then there was the gold: I'd spent the last few years seeking it out and giving advice and support to others who wanted to do the same. I'd always seen myself as a great lover of nature, but now I felt like a fraud and a hypocrite.

The hardest thing was I knew I'd be heading home in a few weeks, and that as soon as I did, I'd slip right back into a life of disposable coffee cups, plastic bags and gas guzzling cars. I told myself I could walk everywhere, bring a bag to the grocery store and not buy take-out coffee, but even if I did, what difference would it really make?

I was aboard the juggernaut and could see well enough there was no stopping it – especially not by a puny former treasure hunter who'd had an epiphany in the jungle. The hope and release I'd felt emerging from the ceremonies was fast turning into a sense of desperate anguish for humanity at large.

*

I was sad to leave the center and the lush cocoon it provided against the outside world - not to mention the friends I'd made there - but I'd received the healing and direction I'd come for and had given my work and expertise in return. It had been a fair exchange, and now it was time to move on. I left Shara in charge of the gardens, and, for as long as he decided to stay, I knew their development would be in good hands.

Shara joined me and Alf in a farewell mapacho ceremony at my tambo the night before I left, where I wished them both well in their respective callings. In my eyes, both men made the world a better place. Even if I did have little faith in their being able to turn the tide of human suffering in any significant way. They thanked me and wished me well in mine also.

"I need to find it first," I reminded them. "Once I've got back and made peace with my family I'll start looking for it, I promise."

Alf and Shara smiled and at each other, as if sharing a private joke at my expense.

"Maybe that is your calling," Alf suggested.

Shara shook his head. "It's the beginning of it. As important as honoring one's self, but it's not enough. I believe that once you've made peace with all of your family your path will become clear, Jason. In fact, I'm sure of it." He smiled and stood and embraced me. "It won't be easy for you, but from what you've told me, nothing ever is. Goodnight, brother, may the maninkari grant you the wisdom to dream well."

Alf stepped forward to hug me, too, then the two of them left me in peace. I stayed on my porch, soaking in the scents and sounds of the jungle. Once I left Iquitos, I knew I'd never come back, and I wanted to savor my final night in the rain forest.

I'd had so many adventures and made so many friends in the Amazon. It had also nearly killed me. First from floods and fever, then at the hands of the men who'd beaten, strangled and thrown me off a cliff.

It was the first time I'd really thought about my attack since the ayahuasca ceremony, when what I'd believed was the spirit of the vine assured me I hadn't brought it upon myself. I realized now that at some point over the past three weeks I'd also stopped blaming my attackers for what had happened; they'd behaved in accordance with their own conditioning, and from a sense of resentment not entirely unjustified. It was strange and surprising and liberating to think of them without the familiar knot of rage twisting my stomach.

More than anything else, the Amazon had brought me healing and emancipation - the depth of which even I hadn't even known I'd needed. I was full of appreciation, awe and respect for this magical land and its inhabitants - the humans, animals and especially the plants. But, in spite of it all, I couldn't wait to leave. It was clear to me now my place was with my family.

Back in Iquitos, I returned to the hostel where I'd met Hedgy and Ben. The city was the same as ever; polluted and overcrowded. But I didn't feel quite as assaulted by it all as I had when I'd first arrived. Even when I discovered I wouldn't be able to get a flight home for a couple weeks, I wasn't too upset. I was still in the strictest phase of the post-sanango dieta and figured I'd have a better chance of sticking to it in Iquitos than I would in the States.

I went to *The Yellow Rose of Texas* and checked out their 'ayahuasca diet' menu options. There was plenty on it I could eat, and I decided on a plate of steamed river fish with rice and vegetables. No seasoning or salt.

When the waitress placed it down in front of me, I felt a big fleshy hand grip my shoulder. For a disorienting instant I mistook the hand for Jim Fletcher's, Anna's rich uncle from Belize. When I looked up into the man's face though, I saw it wasn't him. It was Bob, the big, bald American I'd often seen having breakfast with his equally disgusting logger friend, Ed.

"You sure you want to eat that shit, brother?" he asked me.

"Why not? It's about the healthiest thing on the menu."

He smirked. "Didn't have you down as a health freak the way I seen you putting away your steak and eggs before. You been drinking jungle juice with the hippies and natives and gone all soft in the head or something?"

I held back a quip about it being better than soft in the belly. Up close Bob was huge, with forearms like hams.

"These guys get their fish downriver from my gold operation," he continued. "And you know what that means?"

I glanced at my plate. The iridescent fish-scales took on a sudden metallic hue. "Mercury," I said. "Or do you use cyanide?"

"If I used cyanide they wouldn't have to catch them. They'd be scooping them out of the river as they floated past belly-up. And I'd be demanding a commission for it!" Bob guffawed. "Yeah, we use mercury. You pass the test. I heard you were a gold man, too, but I wanted to make sure. Let me know if you ever need work. I'm always looking for guys who know their stuff to run crews of these dumbass natives for me." He winked. "I'll make it worth your while."

"Thanks for the offer," I said, trying my best to remain civil. "But it doesn't sound like my kind of thing."

"No problem. I get it. I wouldn't work with the little fuckers either if they weren't so damn cheap. See you around, sport." He took his hand off my shoulder and half swaggered, half waddled toward the door. His ruddy head and neck creases glistened in a sheen of sweat by the time he reached it and ducked outside.

"What a dick," I muttered to myself.

"Makes you proud to be an American, doesn't it?" said a guy from the next table with thick, blond dreadlocks tied back from a pale round face. He nodded at my plate. "I'd stick with the rice and veggies though if I were you. Any kind of meat lowers your vibration, whether it contains mercury or not."

"Does it?"

"Yep. One thing I learned from drinking ayahuasca is that eating meat is murder."

"Really? I didn't get that. What about plants?"

"What about them?"

"Isn't that murder, too? I mean, they're living things as well."

Moon-face sighed. "True, but eating fruits and vegetables raises your vibration. Meat lowers it. Meat eaters disgust me. If you'd drank ayahuasca you'd understand."

"I have, and I still don't agree. But I respect your opinion." I gave him a conciliatory smile and tried to return to my dinner.

"Maybe you need to drink again," he suggested. "I'm going to a shaman's house with some friends tomorrow. We've got seven ceremonies planned over the next two weeks. You can join us if you want. It won't cost much."

"Two weeks, huh? That's exactly how long I have left in Iquitos."

My new friend nodded. "That's called synchronicity, man. Ignore that shit at your peril."

The ceremony I'd had with Don Sharaco had been so transformative I couldn't resist taking Moon-face up on his offer. I wanted to wring all the wisdom I could out of the magical brew before heading home.

The ceremonies took place in the living room of a cinder-block house in an Iquitos suburb. The shaman looked the part in his Shipibo costume but he didn't seem to take his calling seriously, preferring to laugh and joke his way through the ceremonies. He only sang the occasional half-hearted icaro, and the ayahuasca he peddled tasted different and was less potent than that of Don Sharaco.

I didn't receive anything very positive or new from these ceremonies and often felt angry and borderline psychotic throughout. Moon-face kept babbling about how amazing everything was while his friends projected an air of privileged awakened hipness. It was all too much for me. I couldn't believe the same medicine could have such dissimilar effects depending on who'd brewed it, and where, and with whom, it was taken.

At the end of the two weeks I'd had enough of the Iquitos

ayahuasca scene to last a lifetime and was readier than ever to return home. I was going to Texas, to stay with my mom and help take care of her miniature horses while I looked for a job and a more permanent place to live. I planned to settle down, save some money and open a plant nursery.

Of course, that wasn't what happened.

Grolars

The hairs on the back of my neck bristled in preternatural warning. Then came the rush of displaced air and the shadow swooping through the twilight three feet above my head. My heart skipped and adrenaline flooded my bloodstream. The eagle flew upward and away, curiosity satisfied, its silhouette shrinking into the fuscia glow of the northeastern sky.

Before my heartrate could return to normal, the vast prehistoric form of a gray whale rose out of the depths in near silence to my right. Close enough to send waves lapping at the portside hull of my dad's Bayliner. The whale glided forward, across the stern of the swaying boat. The rush of air and geyser of salty water sprayed from the behemoth's blowhole startled me anew. With a flip and slap of its monstrous tail, the whale dived and disappeared, leaving me smiling and spattered with water.

"It never gets old, does it?" said my dad, appearing behind me with a cup of coffee in each hand.

I took one of the cups and wrapped my cold fingers around it. "It's definitely beautiful here at this time of year."

My father chuckled. "It's beautiful at any time of year. Just gets a little chilly in the winter is all."

"Not to mention dark. I'll admit it is striking in December. In a stark, icy, snowy kind of way, but blink and the day's gone already and you've missed it."

My dad scratched his grizzled beard and took a sip of his coffee. "That's why you have to savor it all, son. Never know when it'll be gone."

We sat for a while, enjoying the peace and quiet. It was four o'clock in the morning and the sky above us was violet. The northern horizon faded through amethyst and magenta until it resolved in a line of luminous pink over the frosted peaks of the Kenai Mountains. It would stay this way for another hour at least. Summer in Alaska really was magical, especially out on the Kachemak Bay before the sun rises with a day's fishing ahead of us.

"Do I take it from your bitching about the cold you're not going to be coming back to live up here again?"

I shrugged. "I doubt it. I'd like to establish myself somewhere, but where, I don't know yet."

"Would you?"

"Would I what?"

"Like to establish yourself somewhere. Don't go saying things to try make your old man happy. Don't forget, you've got that wild blood in your veins." I glanced at him and he winked. "I should know. You got it from me."

"But you settled down."

"I did. In the end. But it took a while. And I still have to have my fix of the wilds or I'd go crazy. So, what is your long-term plan? Assuming you have one."

I shrugged again. "I'm not sure yet. The last few years have changed me. I got the proverbial bump on the head at about the same time as the literal one. Now I see things differently. And because of that I can't go back to what I was doing before."

"Well, whatever you decide I want you to know it'll be fine by me."

I shot him a look. "You mean you won't be disappointed if I don't get married, buy a house and have kids?"

He laughed at this. "Of course not. You are who you are. It's your life. Don't get me wrong, I'd love to see you married and having kids, but if it's not for you, it's not for you. We have to accept you as you are. If not, we'll all end up resentful and disappointed."

I appreciated my dad's unconditional approval. It gave me a freedom to move forward without the stifling feelings of guilt and regret I'd carried for so long, but barely acknowledged. Surprisingly, my failure to meet my mom's expectations of me had had the same effect.

*

I'd spent six months in Texas, playing a role, badly, before finally accepting I couldn't be the person my mom and the rest of my immediate family wanted me to be. I'd worked on a horse ranch and saved some cash, but apart from that, my Texas sojourn had felt like a waste of time and effort. It hadn't been, but I didn't realize that until after I'd left.

For me, drinking sanango was like installing a new biological operating system. All my memories, skills, strengths and weaknesses were the same, but after the cleanse I could access everything without the glitches, ticks and painful holding patterns which had held me back before. I wanted to make the most of this fresh start, and even though my initial focus had been somewhat narrow, I think it was better to have started that way as it helped reveal a blind spot in my thinking which might have otherwise gone unchecked for too long.

The ayahuasca had worked on me like an x-ray machine. Revealing where I was lacking, broken and damaged; where I continued to mess up and why. But it couldn't really fix anything. Like everyone else, I had to do the healing and repairing part for myself.

Because so many of my visions had centered on me being such a disappointment to my family, I'd initially assumed I should prioritize rectifying their view of me. This was motivated by cowardly, counterproductive vanity and was nothing more than a good excuse to avoid taking larger responsibility for my life; I was concentrating on being a good boy in the eyes of others when I should have been focusing on being a man.

At least that's what Thelma told me.

I'd heard Thelma was lecturing at a local university and I couldn't resist the opportunity to go see her before I left Texas. We went to a campus coffee shop to catch up and I noticed we were soon surrounded by a ring of empty tables. Apart from her hair, which was now iron-gray, Thelma hadn't changed a bit. She was still charming and funny and every bit as terrifying as I remembered. The students, respectful of her viperous aura, knew to keep their distance.

"No wonder you're unhappy, hanging around sweaty horses all day," she observed, a wicked gleam in her eyes. "I couldn't think of anything worse! You need get back to more engaging critters. And as for family, there are many ways of being good to them. In my case it's by avoiding each other in life and keeping them out of my will at death, so they don't have to deal with my house and its contents when I'm gone. Now you of all people would understand what a beautiful gesture that is."

I grinned, thinking of the endless racks of tanks filled with all manner of poisonous spiders, scorpions and other creepy crawlies. Not to mention the multitude of snakes and lizards, most of whom were given free reign of the house. It would take nothing less than one hell of a brave herpetologist to deal with it all when she did pass.

"Jason, you do know you need to stop being a whiny little pussy and man up, right?" Thelma asked, switching tack and pinning me in her steely gaze. "There's more to life than those you were unfortunate enough to be born around. If you can't see eye to eye with the fools, forget about them and move on. There's an old Chinese proverb I want to share with you - 'One who has faced great peril and survived is destined for a life of great happiness, prosperity and purpose'. You heard that one before?"

I shook my head.

"Now, happiness of course is over-rated. Too much happiness is basically mental illness. So, avoid that if you can. Too much prosperity is also a problem. It turns people into shallow assholes with minds full of mush and worry. But purpose - now there's a thing! You survived something that should have killed you dead ten times over. How dare you spend your days making up to mommy and hiding behind stinking horses? You've got gold to work with. And don't think I mean that yellow crap, either." As with most conversations with Thelma, it took a while for me to digest and figure out exactly what she'd been trying to tell me.

*

At around seven in the morning, Mary emerged from the galley with bacon and egg sandwiches and more coffee for my dad and me. Over breakfast, Mary brought me up to date with the latest news from Anchorage, which was mostly local gossip and wider concerns about climate change.

"It's getting harder every year for the Inuits to maintain their way of life," she complained. "More and more are heading into the cities or working out in the oil fields."

Her words made me think of Shara's tribe down in the Amazon. Their struggle was a mirror image of what the northern indigenous were going through. These ancient peoples were being presented with a stark choice: assimilate or die out. What they chose didn't really matter. Either way they would end up being the latest victims of the suicidal steamroller of western civilization.

"In order to survive," Mary continued, "the polar bears are now wandering further than ever in search of food. Speaking of which, you remember your old friend Danny Roach? Well, they say he shot a grolar a few weeks back."

My dad shook his head. "Danny doesn't hunt bears. Must have been someone else."

"Are you sure, Cliff? You know he moved up into the arctic region with Tina and the kids, right?" Mary turned to me. "He's a wonderful father. Not like old Jimmy. A great mechanic, too. Works on helicopters. And his wife's lovely. They've got three girls and a boy."

"Yeah I knew they moved up north," my dad interjected. "And I'm sure the rabbits there have been shaking in fear ever since. But it must have been someone else who shot the grolar. Unless it was on his property."

Mary considered for a moment. "No, you're right. It wasn't Danny. Must have been Gary Rowe. I always get those two mixed up."

I smiled at them both. "What the heck's a grolar?" I asked.

"The grizzly's selfless effort to preserve polar bear DNA," said my dad. "A hybrid. Grizzly father, polar bear mother."

"Or the other way around," said Mary.

"Nope, that would be a pizzly. Father species comes first in these things."

Mary raised her eyebrows. "If you say so, Cliff. But I'm sure they breed both ways. There are second generation cubs now, too."

"I didn't think they were ever found in the same place," I said.

"They didn't used to be," said Mary. "That's what I've been trying to tell you. Less ice means less polar bear hunting grounds and more grizzly-friendly habitat. Not either of their fault, but it means they're now meeting in the middle, and mating!"

"Huh. They're not afraid of each other? They don't fight?"

"I'm sure they squabble, but why should they be afraid? They're all part of the same family after all. Quite literally now. Like your dad said, this could be the only way polar bear genes are going to survive if we keep going the way we are. I think it's great they're getting together. They're making the best of a bad situation, and the results are pretty beautiful and fierce as hell from what I hear."

Something in what Mary said made me think of Shara again. I saw him in my mind's eye, sitting on the porch of my tambo, telling me to make peace with all my family, then calling me brother. I smiled. Man, I could be slow sometimes.

"I guess desperate times call for desperate measures," said my dad, before whistling a few bars of *The Bare Necessities*. "So, Jason, do you at least have any idea where you might be wandering and roaming to after you leave here?"

"Seattle, at first. To visit some friends. After that I'm planning on one of the southern states for winter." I didn't want to tell him which state as I wasn't prepared to answer the inevitable questions about why on earth I'd want to go back to Arkansas.

Although I'd told both my parents about the medicine ceremonies I'd been through in Peru, I hadn't yet told either about the strange drawn-out shamanic exorcism or the visceral visions of the distant past I'd experienced after 'meeting' - while under the influence of the spirit vine - the elderly Native who'd haunted my childhood.

I needed to go back to my old hometown to see if *the ruin* was still there, and to find out more about the tribes who'd first encountered the conquistadors and what had become of them. After doing some online research, I'd discovered the first Europeans to arrive in Arkansas were led by a man named Hernando Soto, and that he'd met with a Native leader called Casqui soon after landing. Reading this gave me goosebumps.

Soto had previously been in Peru and travelled with captive animals from the South American continent, including, presumably, a harpy eagle like the one I'd seen in my vision. Casqui, according to documents written by the Spaniards at the time, had presided over land with advanced integrated agriculture systems - essentially what we now refer to as permaculture. Everything, from the old chief's name to the details of what I could find of his life, fit with my ayahuasca visions. The only thing I had yet to discover was what had happened to Casqui and his people after their initial meetings with Soto and his men.

I was shaken by what I'd discovered. It meant I either had further suppressed memories - this time of some random earlier studies – still lurking in my subconscious, or that Alf had been right all along, and I'd been haunted throughout most of my life by a genuine historical figure. Either way it gave me the shivers and made me want to find out as much as I could about Chief Casqui and Hernando Soto. In the end, though, this journey into the past would have to wait, as my plans for the future were about to supersede it.

*

I left Alaska in late August and flew into a wet and windy Seattle. Hedgy and Ben met me at the airport. Hedgy looked the same as ever, but Ben couldn't have been more different. The pale, skinny drug fiend I'd shared my tambo with in Peru was gone, and in his place was a lean, athletic man with tan skin and bright eyes.

"How you doing, man?" said Ben, gripping my right hand and wrapping his wizard and dwarves and dragon adorned left arm around my shoulders. He stepped back and - in sync with Hedgy - looked me up and down. "Shit, Jason, you look healthy!"

"You do," agreed Hedgy, giving me a hug as well. "Your natural eye color is brown not red. Who knew?"

"Thanks guys. It's so good to see you both."

"You too," they chorused, with matching grins.

"Speaking of healthy, what the heck happened to you?" I asked Ben as we strolled out of the terminal building. "Not that you don't look great as well, Hedgy. But you always do. This guy though...."

Ben grinned. "I got clean, brother. I don't smoke weed, drink beer, snort coke, pop pills, drop acid or shoot the brown anymore. These days I run, hike and mountain bike instead. Seattle's not always rainy, you know."

"Good for you," croaked a bird-like, silver-haired lady in a business suit strutting past. "Be careful you don't take things too far though, son. I gave all that good stuff up in the seventies and now look at me." She glanced back at us and winked. "I've turned into the man!" Chuckling to herself, she stalked away toward an idling blacked-out town car parked at the curb, dragging her cherry-red Louis Vuitton carry-on behind her.

After we watched a uniformed chauffer take the lady's bag and hold the door open for her to climb inside, Ben said, "I don't think I need worry about that. Luckily, I've still got the tattoos holding me back from proper respectability. And I'm drinking ayahuasca every couple of weeks now to keep me in line."

"You are? How?"

He grinned. "Long story. You know how it is. We've got a lot of catching up to do, brother, and all week to do it."

When we got to Ben's place, Hedgy went out to buy some beer for me and her while I helped Ben chop veggies for a stir-fry.

"So, what happened, Ben? Ayahuasca's so good nothing else can compete anymore?"

He laughed. "Not quite. But during my last ceremony in Peru, while you were crawling around in sanango induced blindness, I decided it was time for a change. Simple as that. Crazy thing is my whole genetic code altered once I'd made the decision. I shit you not; the next day I woke up and for the first time in ten years I didn't want to smoke a joint. I stopped craving all the other stuff, too. The ayahuasca had affected my whole physiology and psychology. It was goodbye addictive personality disorder, and hello 'what the fuck do I do instead?'. It was a trip, man, I'm telling you."

"Sounds like it. So what's the deal with the ayahuasca here? You importing coke bottles from the toothless dude in Iquitos?"

Ben laughed. "Nah, my dealing days are done now as well. Believe it or not, I've joined a spiritual group. Ayahuasca's their sacrament. Mine too, now."

"Sounds a bit churchy."

"It is. But not in a bad way. Although it is based on a kind of Catholicism."

I made a face and Ben held up his hand.

"Here me out, man. Don't judge when you can understand. The guy who started the group was a Brazilian rubber tapper, back in the sixties. The only spiritual instruction he'd ever had was from the Catholics, so it's understandable he drew on their vocabulary and archetypes to describe his experiences and ideas. Anyway, it's a solid group. Real people drinking ayahuasca to better themselves in order to better the planet. Isn't that what it should be about?"

"Of course. But I don't know, man; the religious stuff would leave me cold. Besides, I did a few ceremonies in Iquitos after I left the center and they were all over the place. I think having a proper shaman is important."

"Agreed. That's what's so cool about these guys. You can't just walk in, pay your money and drink. You have to get to know them and they have to get to know you. Afterward you'll either be invited to drink the tea with them or not. Kind of like joining a tribe where everyone is responsible for everyone else's wellbeing. The shaman role is shared within a circle of elders, so there's no danger of guru complex and abuses arising. Seriously, man, even though it's mostly middle-class, middle-aged white people with a sprinkling of Brazilian ex-pats, it feels more genuine and relevant than anything I experienced in Peru. And we were lucky with where we drank, too."

Hedgy arrived while Ben had been mid-flow. She handed me a beer and sat up on the counter between us. "Sorry to interrupt your proselytizing, Ben. But a little more chopping and cooking and a lot less talking would be better. I'm starving! Oh, by the way, I saw Daniela and Cayetana outside and invited them to join us for breakfast tomorrow."

"Cool. Where?"

"Here, of course. They're your neighbors for Christ's sake. And anyway, I'm planning on getting too drunk to drive home and Jason's crashing here, so it makes sense to do it here. Plus, you won't be nursing a hangover so you'll be able to cook for us all again." She reached for a carrot stick and crunched it between her teeth. "Aren't you glad you're such a better man these days?"

Ben handed her a knife. "Shut up and chop the mushrooms!"

We stayed up late, reminiscing about Peru, making fun of Tori and Yusuf and each other while putting the world to rights. At around two in the morning I went to the spare room and fell into a deep sleep, waking some time later to drink the glass of water Ben had left on my bedside table. After that I lay awake for a while, listening to the rain and the branches lashing against the window pane.

I was soon lulled back to sleep and dreamed I was in the jungle again. This time there were no vines. Instead, walking upon one side of me was a jaguar, and on the other a faceless woman, whom I felt I knew well but couldn't quite place. Overhead, monkeys and parrots were playing an elaborate game with each other, squawking and screeching back and forth. I found I could understood their banter and laughed along with them. The woman spoke to me, telepathically, telling me she was happy I was back. The jaguar told me the same thing and that I belonged there. Understanding their thoughts was as natural to me as blinking.

The light was gentle and the leaves smelled of lime and mint and the mossy ground was soft beneath my bare feet. Something I saw as innocence and slow-learned wisdom swirled like pollen around my head. I heard the whispered echo of a promise; my own promise to the guayusa tree and the vines from the bottom of the cliff. I was at peace and in balance. I'd survived and healed and was ready to run. As soon as I thought of it, we all took off sprinting together - the woman, the jaguar and me. Above, the air was buffeted by snapping wings and the violent rustle of a monkey on the move through the branches. My world became a blur of jade and golden light, awash with endless possibility and my chest swelled and almost burst with hope.

*

I awoke to the sound of a door opening and voices speaking in Spanish. It was disorienting and for a moment I had no clue where I was. The thick foliage scratching up against the window didn't help, but the dismal sky beyond did. I was still in Seattle. I recognized one of the voices now as Ben, whom I recalled was fluent in Spanish. I could distinguish two more, both female; Daniela and Cayetana coming for breakfast, I presumed.

By the time I'd showered and dressed, everyone was already sat around the kitchen table munching their way through an enormous stack of pancakes and plates full of bacon and fried eggs.

"You see, Jason," said Hedgy, when she spotted me skulking in the doorway. "I told you Ben would make someone a wonderful wife one day. Look at this spread! Come take a seat and I'll introduce you to the girls."

Cayetana and Daniela were Ecuadorian exchange students in their early twenties. Cayetana was studying music. Daniela was working on a PhD in environmental studies. Both were passionate about protecting the rain forest and were members of various activist groups in Ecuador and the States. We had a lot to talk about.

"So, you're the permaculture guy?" Daniela asked after we'd been chatting for a while. "I remember Hedgy telling me about you now. So, you're going to feed the world for us, huh?"

"If only."

"You have to according to the United Nations. They published a study last September saying the only way to combat an imminent worldwide food crisis is with small-scale, organic farming. Monocropping and chemicals are out if we're going to survive as a species. You didn't hear about it?"

"Not until now. How did Monsanto and the other crooks at the top of the food chain take the news?"

"They pretty much buried it. I mean, it's your field and even you didn't get to hear about it."

"Well, I was in Texas at the time."

"And Texas is under a rock? Come on, guy. You should be out there making sure we're all fed. These pancakes are not going to last forever."

"Damn right," said Hedgy, heaping five more onto her plate and dumping the syrup over them.

Daniela shook her head. "Ah, the American way. So, Jason, would you ever go back to my country?"

I shrugged. "Sure. Why not?"

"I don't know, maybe because you nearly got killed there. Wouldn't you be too afraid or angry to go?"

I smiled. "Not anymore. Well, I might be a little afraid, depending on where I'd be going exactly, but not enough to stop me. Why?"

"Because I'm working with a tribe up near the Colombian border, trying to raise awareness of what is happing there so we can help them. Their lands have been destroyed by Texaco and Chevron and now the national oil company. The president said mother nature should have rights, and even wrote it into the constitution, but it's all bullshit. He's selling up mining rights in every part of the country, including the national parks, and is still drilling and spilling oil all over the rain forest. Nobody can do anything about it as he's outlawed all forms of activism within the country. With what I am doing here even I run the risk of being arrested when I return."

"Shit. I had no idea things were so bad. But, with respect, what does it have to do with me?"

"Depends. We inherited this messed. It's not our fault, but it will be if we do nothing. People talk about *the* environment like it's something separate from us, like it's some abstract political issue. But it's not. It's *our* environment. We all live here. The tribes I'm telling you about are starving. Their hunting grounds are poisoned. The soil is bad. I don't know, if you want to maybe you can help them fix it. Or at least enough so they can grow something to eat."

Everyone at the table had fallen silent. All eyes were on me. I shrugged. "I might be able to do something."

*

The plane touched down in Quito in the pre-dawn darkness. I passed through customs and immigration, and by the time I got outside it was a bright, clear morning. Mauro, a young member of the tribe Daniela had described, met me and led me out to a battered pick-up truck. He threw my bags in the back, then hustled around to the driver's door.

"Let's go. It's a long journey," he said, climbing in.

And that was it. The course of my life had changed yet again.

Twelve hours later I was back in the jungle. Not for riches, or adventure, or personal healing. I was done with taking. It was time I started giving.

I haven't looked back since.

Epilogue & Author's Notes

Jason Pednault   Matheu DeSilva

To Fly

Jason Pednault contacted me in 2015 to ask if I'd be interested in ghost-writing an account of his survival story in the Ecuadorian Amazon.

I was reluctant at first because Jason and I already knew each other. We'd met just after his failed expedition to the Jatanyacu, when he and his companions were flooded out and almost washed away by the river.

Around this time, the locals of Coaque had invited him to return to the coast of Ecuador to investigate some remote Incan burial mounds, and he was looking for someone to oversee the delicate task of transporting any artifacts recovered. A mutual friend who knew of my previous work with treasure hunters and smugglers in southern Asia (that's another story, perhaps even another book!), recommended me for the job.

On first meeting, Jason seemed like an amiable and capable, if somewhat geeky, redneck. I liked him, and was impressed by his drive and attention to detail. He was a doer, not a talker, and the adventure he presented was very tempting.

In the end, I decided to pass on his offer. I'd left that kind of life behind and was content to be working full-time on writing and music projects. I'd also noticed Jason had a tenacious, unsettled and frenetic air that was all too familiar to me. I imagined he would either amass a fortune from his wild expeditions or die trying. Maybe both.

Sometime later, when I was told of his attack, I wasn't at all surprised to hear he'd gone straight back to gold prospecting afterward. This idea that Jason - like so many others - seemed incapable of any real change was why I had such little initial interest in writing his story.

I couldn't have been happier to have been so wrong about somebody.

When we met to discuss his idea for a book, he explained how he had now given up prospecting and taken to a life of hands-on activism. He was working with indigenous Peruvians and Ecuadorians, campaigning to protect their lands from exploitation, and teaching integrated farming techniques to enable them to feed themselves while still preserving the rainforest in which they lived.

He'd also launched a cooperative business called *Jungle Visions* (previously *Pure Synergy Herbs*), selling wild-harvested products and artwork direct from the Amazon basin. *Jungle Visions* was conceived to empower indigenous communities to be able to preserve their traditional ways of life while still earning money for their labors, should they desire, by providing a real and practical alternative to illegal logging or working for oil or mining companies. It's worth mentioning that through *Jungle Visions,* Jason is now directly helping the community where the men who tried to kill him live.

His fight against 'Big Oil' also recently brought him back to the States, where he raised money for vital supplies and journeyed to stand in solidarity with the water protectors during the brutal winter at Standing Rock.

As impactful, exciting and mysterious as Jason's survival from the attack on his life was, it was this transformation from gold prospector to activist that convinced me his was a tale needing to be told.

Once we began the interviews that provided the raw material for the book, another, deeper, story emerged - that of his troubled, haunted childhood, his time in Presa River, and how they tied in to the cleansing and renewal, and eventual repurposing, he later experienced from the medicine ceremonies in the Peruvian Amazon.

This story framed his journey in a larger context - that of the relationships between different cultures; the misunderstandings, fear and (in)human corruption which have often resulted in the annihilation of those who live closest to nature by those who wish to subdue and control it. Each time one of these tribes or cultures is lost - whether driven into extinction or assimilated into the homogenous gloop - we all lose a wealth of valuable information and hard-earned wisdom.

On the day we recorded the details of Jason's attack, we found the body of a fledgling crow on the veranda outside the room we were using for the interviews. We could see the nest from which it fell in a forked branch overhead. Jason, the consummate climber and nature lover, scaled the tree to take a careful peek inside the nest. All the hatchlings were gone. Judging by the remnants of their eggs, there had been four of them in total; three survivors and one casualty.

We wondered why the bird on the veranda hadn't made it. Then we considered why its siblings had; imagining the fledgling birds utilizing a combination of genetic memory, instinct, observation, learning - perhaps along with some cawed maternal instruction - as they practiced, hopping and fluttering and stretching their wings, contemplating the big leap. Whatever the method, there came a time when they had to be brave and risk falling in order to fly.

The similarities between this process and what Jason had gone through at the bottom of the cliff after his attack were startling. He'd had to use all his skill and instincts, and no little amount of bravery, to survive.

But he hadn't done it alone. He'd received help in the form of the vines and the guayusa tree, the rocks and scraggly shrubs of the cliff-face. As a result, just as a bird learns to respect the flow of the air currents keeping them aloft, Jason recognized the need to nurture a stronger relationship with what had saved him, and which continues, under extreme duress, to sustain us all: our environment.

Within the Amazon basin, along with other remote places across the planet, there is still a chance to protect and preserve vital knowledge and resources; the fauna and flora of the wild. There is also an opportunity to combine those resources and knowledge with our advanced technologies in responsible and enlightened ways. To not only restore our symbiotic relationship with the world in which we live, but to allow it to evolve in more conscious and sustainable ways.

But will we learn from the mistakes of our past and take that chance?

Perhaps, like Jason, we should. Because the human race is falling fast. We're heading the way of the dead crow because too many of us have allowed ourselves to be tricked into believing we can no longer dream new dreams - the wings of our imagination have been clipped by the virus of rabid acquisition (as described by Chief Casqui in Jason's ayahuasca vision).

Right now we're destroying what we need to live: our waterways, the forests that clean our air, the soil that sustains our crops, the plants and animals that maintain our ecosystems, the individual creativity and collective consciousness that feeds our souls and informs our humanity.

In short, we're losing the art of living.

If we're to survive as a species, not to mention increase our quality of life, each of us has a personal responsibility to rectify that in any way we can.

The question is simple:

Do we want to fall through life until we splat, taking our future generations with us, or do we want to realize our potential and soar?

Jason Pednault   Matheu DeSilva

Timelines and characterizations:

The timeline of certain events have been slightly altered in order to enhance the flow of the story.

The characters, with the exception of Jason and his immediate family members, are based upon amalgamations of real people and are therefore not to be considered a full portrait of any actual persons, living or dead.

Concerning ayahuasca:

Like Jason, my personal experiences with ayahuasca and other psycho-tropic substances have been very positive and beneficial (if somewhat challenging on occasion).

However, we are aware this is not the case for everyone.

Jason's friend, Leslie Allison, to whom this book is dedicated, lost her life during an ayahuasca ceremony in Ecuador. The details of exactly how this happened are still a mystery. We do know her death was caused by severe physical trauma sustained during the evening of the ceremony and not as a direct result of drinking ayahuasca itself. There were over forty participants at the ceremony and only three facilitators.

If you, as an individual, make the decision to explore these powerful plant medicines further, we advise that you observe due diligence, respect and caution.

Be wise and take care.

Jason Pednault   Matheu DeSilva

Falling to Fly

Jason Pednault   Matheu DeSilva

Falling to Fly

Jason Pednault   Matheu DeSilva

Made in the USA
San Bernardino, CA
23 July 2018